SPACE

VISUAL ENCYCLOPEDIA

HEATHER COUPER AND NIGEL HENBEST

CONTENTS

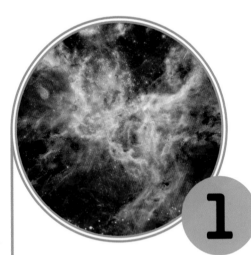

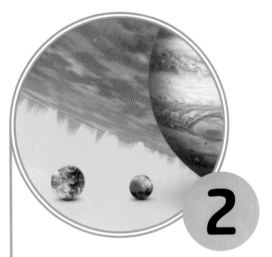

1

2

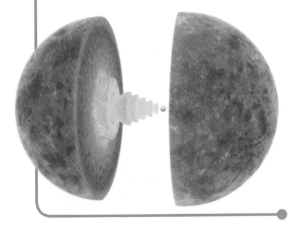

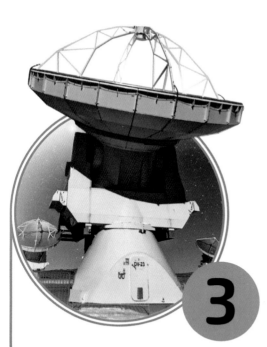

4

SPACE EXPLORATION 94

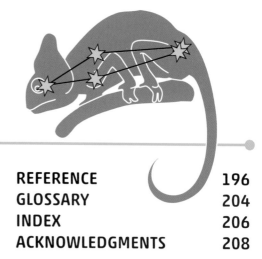

SCALES AND SIZES

This book contains profiles of planets next to a scale drawing of Earth to indicate their size.

Earth

12,756 km
(7,926 miles)

1
WHAT IS SPACE?

Look into the sky on a clear, dark night, and you can see thousands of stars floating in the vastness of space. These glowing balls of hot gas are born inside large clouds of dust and gas called nebulae, and form giant star-cities called galaxies. Many stars are orbited by planets – balls of rock, ice, or gas. The night sky that you can see is just a tiny part of our immense Universe.

THE BIG BANG

The whole Universe started out **smaller than an atom**.

In the beginning, there was nothing. No stars or planets; no space or time. Then a tiny point of energy appeared. Time began to tick. An instant later, that tiny blob blew up in a vast explosion called the Big Bang that created all the matter in a Universe that continues to expand.

Energy turns into particles

Atoms form

1 2 3 4 5 6 7

One millionth of a second after the Big Bang

350,000 years after the Big Bang

400,000 years after the Big Bang

The Universe becomes transparent as light begins to pass freely through space

1 A tiny, hot point of energy, smaller than an atom, suddenly appears: astronomers still do not know why the Big Bang happened.

2 In a fraction of a second, the Universe explodes. This period, called "inflation", produces vast amounts of heat energy.

3 At a temperature of trillions of trillions of degrees, the intense heat creates matter and its opposite, antimatter.

4 Matter and antimatter destroy each other. The little matter left over begins to form protons and neutrons.

5 After 380,000 years, the Universe has cooled enough for atomic nuclei to combine with electrons to make hydrogen and helium atoms.

Heat radiation from the Big Bang fills the Universe. Astronomers map it with radio telescopes: the best view has come from the Planck space telescope orbiting Earth. This image shows small differences in temperature in the radiation, with red being the hottest.

8

9

10

Solar System forms

6 Swirling clouds of hydrogen and helium gas are pulled into dense clumps by the gravity of mysterious dark matter.

7 After 500 million years, the first stars are born in galaxies centred on quasars (hot gas clouds around massive black holes).

8 After 5 billion years, the Universe consists of vast clusters of galaxies arranged in threads with gigantic voids between them.

9 Some 8 billion years after the Big Bang, the expansion of the Universe begins to accelerate, driven by dark energy.

10 Now 13.8 billion years old, the Universe grows ever larger, but will eventually fade away as stars die.

THE UNIVERSE

What is space?

Observable Universe
Astronomers have looked 13.8 billion light years into the Universe.

Virgo supercluster
Our galaxy is just one of hundreds of thousands of galaxies that are clustered together in a group called the Virgo supercluster.

Milky Way
The Sun and its neighbouring stars occupy a tiny fraction of the Milky Way galaxy, a vast, swirling disc that contains about 400 billion stars.

The Universe includes everything that exists. Its scale defies imagination. Even travelling at light-speed, it would take billions of years to cross the part of the Universe we can see. Matter in the Universe is clumped into galaxies that float together in giant groups called superclusters. Our Milky Way, which contains the Sun and Earth, is a typical galaxy.

We can only see out as far as 13.8 billion light years, but the Universe probably extends much further

The Universe contains **more stars than sand grains** on Earth's beaches.

Stars make up the Milky Way, with some gas and dust thinly spread between them

The Sun is a fairly typical star. Many stars probably have a family of planets, like our Solar System.

Local stars
The nearest star to the Sun is about 4.2 light years away.

Solar System
Earth belongs to a family of other planets and rocky objects that orbit the Sun.

Earth
Our world is a small, rocky planet travelling through the vastness of space.

DARK MATTER AND DARK ENERGY

One of the most amazing things that scientists have discovered is that most of the Universe is dark and invisible. Galaxies are held together by the gravity of dark matter, probably made of tiny subatomic particles filling space. On a larger scale, a mysterious force called dark energy is ripping the entire Universe apart.

Nearby star in our Milky Way

DATA FILE

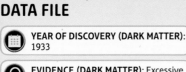

YEAR OF DISCOVERY (DARK MATTER): 1933

EVIDENCE (DARK MATTER): Excessive power of gravity controlling galaxies

YEAR OF DISCOVERY (DARK ENERGY): 1998

EVIDENCE (DARK ENERGY): Acceleration of the Universe's expansion

One of the hundreds of galaxies making up this galaxy cluster

Pink region shows a cloud of hot gas that envelopes a galaxy cluster

Four galaxy clusters collided to create the giant cluster seen in this combination of X-ray and visible light images. Dark matter threads its way through the cluster, as seen in blue.

Dark matter is everywhere – even **inside your body**.

Blue region is filled with dark matter, whose gravity is confining the galaxies

INGREDIENTS OF THE UNIVERSE

Ordinary matter – including the atoms that make up stars, planets, and ourselves – is only a minor ingredient in the Universe. It is far outweighed by invisible dark matter. But the most important constituent is dark energy, whose power is accelerating the expansion of the Universe.

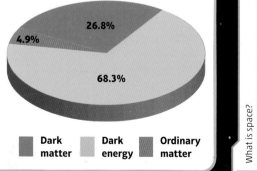

26.8%

4.9%

68.3%

◼ Dark matter ◼ Dark energy ◼ Ordinary matter

GALAXIES

Families of stars called galaxies, of all shapes and sizes, throng the Universe. Each galaxy is a giant star-city, composed of billions of stars, and often spectacular clouds of gas and dust as well. The Sun is just one of 400 billion stars making up our home galaxy, the Milky Way.

The core of a spiral galaxy consists of old stars

DATA FILE

GALAXIES IN OBSERVABLE UNIVERSE: More than 220 billion

SMALLEST KNOWN GALAXY: Segue 2, 200 light years across

LARGEST KNOWN GALAXY: IC 1101, about 6 million light years across

CLOSEST GALAXY: Canis Major Dwarf Galaxy, 42,000 light years away

Molecular clouds are dense patches of gas and dust

Bright regions are areas of starbirth

Winding lanes made of billions of stars form the colossal arms of the Whirlpool Galaxy, which sweeps through space like a sparkling cosmic wheel.

Giant galaxies are cannibals: they have grown large by **swallowing their neighbours**.

GALAXY SHAPES

Galaxies come in four main shapes. The most beautiful galaxies are the spirals, in which the stars lie in curved arms. In the barred spirals, the central stars form a short straight "bar". Elliptical galaxies are round balls of old stars, with little gas and dust. Shapeless galaxies are called irregular.

Spiral

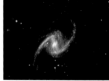

Barred spiral

Elliptical

Irregular (detail)

What is space?

STARBIRTH

Stars are born in vast clouds of interstellar gas and dust called nebulae. Astronomers can see them because of the infrared (heat) radiation and radio waves stars produce when they form. The energetic young stars eventually emerge from the gloom and light up the surrounding cloud.

Dark lanes of dust are the remains of the cloud from which the stars were born

The Trapezium is a clutch of young stars, born only 300,000 years ago

DATA FILE

NUMBER OF NEBULAE IN MILKY WAY: 8,000

SMALLEST NEBULA: G75.78+0.34, 0.05 light years across

LARGEST NEBULA: NGC 604, 1,500 light years across

CLOSEST NEBULA: Horsehead Nebula, 1,250 light years away

One of the brightest nebulae in the night sky, the Orion Nebula is a major star-forming region, lying at a distance of 1,340 light years. It is easily visible through binoculars in the night sky, although it will not be as bright as seen here.

Hydrogen gas glows red as it is heated by the stars in the Trapezium

HOW STARS FORM

Gravity plays a vital role in starbirth, pulling together the gas and dust in a galaxy until it condenses into a shining star.

1 Starbirth begins in a dense molecular cloud, rich in molecules such as hydrogen and carbon monoxide.

2 Gravity causes the cloud's centre to collapse into denser fragments, which will form a cluster of stars.

3 The shrinking gas in each fragment grows hot enough to emit infrared rays: it is now a "protostar".

4 Spinning gas and dust around the protostar form a thin disc: hot gas streams out from its poles.

5 The centre of the protostar becomes hot enough to ignite nuclear reactions: a star is born.

Three new stars are born in the Milky Way every year.

TYPES OF STAR

Most stars – including our Sun – are bright, hot balls of gas that are powered by nuclear reactions deep inside their core. Our Sun is quite an average, middle-of-the-road type – known as a main sequence star – but others encompass unbelievable extremes of size, brightness, and temperature. The more massive a star, the hotter and brighter it glows.

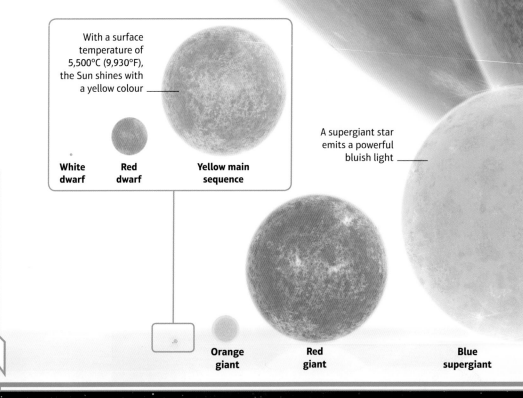

Red supergiant

With a surface temperature of 5,500°C (9,930°F), the Sun shines with a yellow colour

A supergiant star emits a powerful bluish light

White dwarf

Red dwarf

Yellow main sequence

Orange giant

Red giant

Blue supergiant

STAR CLUSTERS

A giant ball of more than a million stars, the globular cluster NGC 2808 is around 12.5 billion years old, about the same age as our Milky Way.

The Butterfly Cluster, in the constellation Scorpius, contains less than 100 stars. It is one of the largest and brightest clusters in the Milky Way.

You could **fit five billion Suns** inside the biggest star, **UY Scuti**.

DATA FILE

 SMALLEST DWARF STAR: Sirius B, white dwarf, 11,700 km (7,270 miles) across

 LARGEST STAR: UY Scuti, red supergiant, 2.4 billion km (1.5 billion miles) across

BRIGHTEST STAR: R136a1, 9 million times brighter than the Sun

 CLOSEST STAR TO THE SUN: Proxima Centauri, 4.24 light years away

Blue hypergiant

LIFE OF A STAR

Lightweight stars live the longest – up to **10 trillion years**.

For most of its life, a star is powered by the nuclear fusion of hydrogen into helium. When all the hydrogen is used up, a lightweight star puffs off its outer layers, as a glowing planetary nebula. However, a heavier star can fuse a whole sequence of elements, turning into a red supergiant before exploding as a supernova.

DATA FILE

LIGHTEST STAR: GJ 1245C, 0.07 of the Sun's weight

HEAVIEST STAR: R136a1, 315 times heavier than the Sun

COOLEST STAR: 2MASS J0523-1403, 1,800°C (3,270°F) at the surface

HOTTEST STAR: WR 102, 210,000°C (378,030°F) at the surface

At an age of less than 10 million years, the star Betelgeuse is nearing the end of its life.

Some scientists think there may be a planet around Betelgeuse

Betelgeuse lies around 640 light years away from Earth in the constellation of Orion

LIFE OF A HEAVYWEIGHT STAR

The heaviest stars have the hottest cores, and burn the fuel in their nucleus so quickly they die after only 10 million years. The reactions take place in a series of stages.

1 Soon after forming, a star begins a long period in which it hardly changes, as hydrogen is converted to helium in its core.

2 When the hydrogen starts to run out, changes in the core create extra energy and cause the star's outer layers to swell and form a red supergiant.

3 The supergiant explodes as a supernova and blows off its outer layers, leaving behind only its core, which begins to shrink. It will then become either a neutron star or a black hole.

SUPERNOVAE

The brightest supernova, ASASSN-15lh, **outshone the Sun 570 billion times.**

A lightweight star like the Sun dies quietly, puffing off its outer layers in a glowing cloud (a planetary nebula). But a heavyweight star suffers a spectacular death. As its iron core collapses, numerous tiny particles, called neutrinos, flood out, blasting the star apart in a dazzling supernova explosion.

DATA FILE

FIRST RECORDED: 185 CE, by Chinese astronomers

MOST RECENT NAKED-EYE SUPERNOVA: SN1987A, in 1987

NEAREST OBSERVED SUPERNOVA: SN1054 (created the Crab Nebula)

MOST PROLIFIC DISCOVERER: Tom Boles, UK, 155 supernovae

PULSARS

The core of a supernova may collapse to become a tiny neutron star, an object so dense that a pinhead of its matter would weigh a million tonnes. Its magnetism beams energy outwards, like a lighthouse, creating regular pulses as the star spins – hence the alternative name, "pulsar".

Beam of energy

A pulsar is the size of a city, but contains as much matter as the Sun

DATA FILE

FIRST DISCOVERED: 1967, by British astronomer Jocelyn Bell (PSR1919+21)

FASTEST SPINNING: PSR J1748-2446ad (716 rotations per second)

SLOWEST SPINNING: SXP 1062 (1 rotation every 18 minutes)

CLOSEST PULSAR: Geminga (815 light years)

The first pulsar was jokingly called LGM1 – **"Little Green Men 1".**

BLACK HOLES

When a red supergiant dies, its core collapses under its own weight, forming a black hole. It is "black" because its intense gravity prevents light escaping, and a "hole" because it traps everything else too. The centres of galaxies often contain supermassive black holes, billions of times heavier than the Sun.

SPAGHETTIFICATION

Gravity strongest on legs, stretching them first

Black hole

The gravitational pull of a black hole rises so steeply nearby that an astronaut falling into one would be stretched like spaghetti and torn apart.

A brilliant quasar is **hot gas around a supermassive black hole**.

Ring of light bent around
from behind by gravity

Event horizon is the edge of
the black hole; anything that
goes beyond this point
cannot escape

The singularity – the central
part of a black hole – contains
the crushed remains of a star

Accumulated cloud of
gas, called an accretion
disc, swirls inwards

DATA FILE

FIRST DISCOVERED: 1971
(Cygnus X-1)

LARGEST BLACK HOLE: H1821+643,
30 billion times heavier than the Sun

CLOSEST BLACK HOLE TO SUN: V616
Monocerotis, 2,800 light years away

EVIDENCE: Gas clouds or stars rapidly
orbiting something invisible

What is space?

PLANETARY SYSTEMS

Our planet Earth and its seven sister planets make up the Solar System. For centuries it was the only planetary system known to astronomers. In the past couple of decades, though, they have found hundreds of planetary systems around other stars. They have all formed from the whirling disc of gas and dust that surrounded their parent star when it was still a protostar – a star still in its early stage of formation.

About 117 light years away from us, the star Kepler-444 is similar to the Sun in its size and brightness, but is more than twice as old

DATA FILE

NUMBER OF KNOWN STARS WITH AT LEAST ONE PLANET: 2,571

NUMBER OF MULTIPLE PLANET SYSTEMS: 585

SYSTEM WITH MOST PLANETS: HD 10180, 7 (confirmed), possibly 9

CLOSEST PLANETARY SYSTEM: Gliese 876, 4 planets, 15 light years away

HOW THE SUN'S PLANETS FORMED

Dust specks orbiting the young Sun stuck together to form pebbles. These gathered into worlds the size of our Moon (called "planetesimals"), which then collided to build up the rocky planets, such as Earth. Further out, Jupiter and Saturn scooped up gas, and grew into two of the four gas giants.

Most planetary systems are **small enough to fit inside Earth's orbit** around the Sun.

The star's planets orbit it so closely that they are too hot for life to exist there

Formed over 11 billion years ago, the Kepler-444 system is home to five planets smaller than Earth. The planets in this compact system orbit their Sun-like star in less than ten days.

What is space?

2 OUR SOLAR SYSTEM

The small patch of the Universe we understand best is the "Solar System" – the Sun and its family of planets. These worlds include our planet Earth, the third rock from the Sun, and the gas giants beyond. Earth is the only planet in the Universe known to support life.

THE SUN'S FAMILY

The Solar System is our home in space. This is the Sun's family, a community of eight planets, plus millions of tiny bodies, such as comets and asteroids. The planets are divided into two groups: small, rocky worlds like Earth, which orbit close to the Sun; and, further out, the "gas giants", which are huge bodies cloaked in vast atmospheres of gas.

Our Solar System is **4.6 billion years old** – young on the cosmic scale.

1 THE SUN

2 MERCURY
Distance from the Sun:
57.8 million km (36 million miles)

3 VENUS
Distance from the Sun:
108.2 million km (67 million miles)

4 EARTH
Distance from the Sun:
149.6 million km (93 million miles)

5 THE MOON
Distance from Earth:
385,000 km (239,220 miles)

6 MARS
Distance from the Sun:
227.9 million km (141.6 million miles)

7 THE ASTEROID BELT
Distance from the Sun:
270–620 million km (168–385 million miles)

8 JUPITER
Distance from the Sun:
778 million km (484 million miles)

9 SATURN
Distance from the Sun:
1.4 billion km (890 million miles)

10 URANUS
Distance from the Sun:
2.8 billion km (1.7 billion miles)

11 NEPTUNE
Distance from the Sun:
4.5 billion km (2.8 billion miles)

Planetary orbit around the Sun

Held together by the Sun's gravity, the eight planets orbit the central star along oval paths. The greater the distance from the Sun, the longer it takes for a planet to complete its orbit.

THE SUN

Our local star, the Sun, is the hub of the Solar System. Because it is so massive, it holds the planets in orbit with its mighty gravity. It is also a colossal nuclear reactor. At its core, which sears at 15 million °C (59 million °F), the Sun fuses hydrogen into helium, creating the heat and light that makes life possible on Earth.

The core, the powerhouse of the Sun, is 10 times denser than gold

DATA FILE

DIAMETER: 1.4 million km (864,900 miles)

TYPE: Yellow main sequence

SURFACE TEMPERATURE: 5,500°C (9,930°F)

DISTANCE FROM EARTH: 149.6 million km (93 million miles)

Earth (for scale)

The Sun's photosphere is its visible surface. It is made entirely of gas.

SUNSPOTS

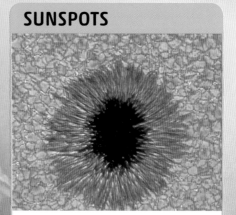

Roughly every 11 years, the Sun breaks out in spots. These "sunspots" are cooler areas in the Sun's atmosphere, in which its gases are trapped by its powerful magnetic field. They can cause explosions, hurling electrically charged particles towards Earth, which can sometimes affect radio communications.

The Sun converts **four million tonnes** of its matter into energy every second.

Prominences are giant arches of hot gas. They follow loops in the Sun's invisible magnetic field.

MERCURY

The smallest planet in the Solar System, Mercury is the closest world to the Sun. Like our Moon, it is heavily cratered, a result of impacts by asteroids in the past. This mini-world also seems to have shrunk, like a dried-up apple. As a result, the planet is crisscrossed with "wrinkle ridges". Being so close to the Sun, Mercury is difficult to spot in the sky.

Mercury's year is **shorter than its day**.

Craters on Mercury, such as the Mendelssohn Crater, are often named after composers and artists

IMPACT BASIN

At 1,300 km (800 miles) across, the Caloris Basin is the biggest of Mercury's scars. It was gouged out by a collision with a giant asteroid 3.8 billion years ago. This coloured image shows the lava floods that filled the basin in orange, and later craters in blue.

Streaks of pale material blasted out by asteroid impacts surround large craters

Mercury's huge iron core is 3,600 km (2,235 miles) wide, making up the bulk of the planet

The planet's mantle is composed of silicate rocks, and is far less dense than its core

Mercury's surface is a single, solid shell, unlike Earth's crust, which is made of moving plates

Mercury has a very thin atmosphere, most of which has been blown away by the solar wind

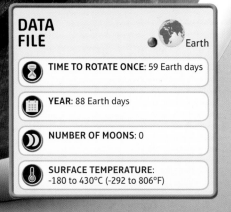

DATA FILE

Earth

⌛ **TIME TO ROTATE ONCE**: 59 Earth days

📅 **YEAR**: 88 Earth days

🌙 **NUMBER OF MOONS**: 0

🌡 **SURFACE TEMPERATURE**: -180 to 430°C (-292 to 806°F)

VENUS

Venus is Earth's twin in size, but that is where the similarity ends. Unlike our life-friendly planet, Venus is covered in a thick layer of cloud that traps the heat, caused by active volcanoes. Its temperature is hotter than an oven, its carbon-dioxide atmosphere is laced with sulfuric acid, and the air pressure is 90 times that on Earth. If you went to Venus, you would be roasted, corroded, suffocated, and squashed.

The core of Venus is made of iron

Using radar, astronomers have mapped the parched, volcanic surface of Venus

The Dali Chasma is a 2,000 km (1,243 mile) long network of long canyons and valleys

The mantle is made of soft rock warmed by the core, which slowly churns about over millions of years

Venus is so hot, rocks on its surface **glow in the dark**.

The planet's crust is made of basalt and other silicate rocks

HIGHEST MOUNTAIN

Venus is the only planet named after a woman, as are most of its features. However, the planet's highest mountain, Maxwell Montes, is the only feature on Venus to be named after a man. It commemorates James Clerk Maxwell, a Scottish physicist who predicted radio waves. The mountain stands at a height of 11 km (7 miles).

Thick clouds cloak the entire planet

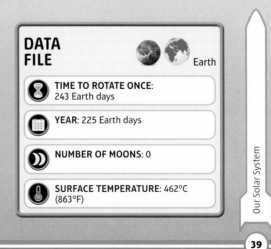

DATA FILE

Earth

TIME TO ROTATE ONCE:
243 Earth days

YEAR: 225 Earth days

NUMBER OF MOONS: 0

SURFACE TEMPERATURE: 462°C (863°F)

EARTH

Our world is special. At 149.6 million km (93 million miles) from the Sun, it is the only planet boasting huge oceans of liquid water, rivers, and seas. Water is essential for life; and life, in all its forms, teems on Earth. Life on Earth began soon after our planet had cooled from the hot bombardment from asteroids and meteorites.

After 4.6 billion years, **the plates in Earth's crust still shift**, causing earthquakes.

Earth's atmosphere is made up of nitrogen and the essential gas oxygen, which we need in order to breathe

Earth's oceans contain about 97 per cent of the planet's water

At 5,500°C (9,930°F), Earth's inner core is made of a solid iron-nickel alloy

LIFE-GIVING LIQUID

Water drives life. It helps to support the chemical reactions that enable life to thrive: our bodies are made up of 60 per cent water. Water is rare on the other planets, which are either too hot or too cold – it is only in the "Goldilocks Zone" (where the temperature is neither too hot nor too cold) around our Sun that liquid water can exist.

The outer core is liquid, and is believed to generate Earth's powerful magnetic field

The mantle is the thickest layer in our planet's interior. It churns about slowly, causing the rigid crust above to move.

Earth's constantly-shifting crust is home to violent earthquakes and destructive volcanoes

DATA FILE

TIME TO ROTATE ONCE: 23.9 hours

YEAR: 365.26 days

NUMBER OF MOONS: 1

SURFACE TEMPERATURE: -93 to 71°C (-135 to 160°F)

THE MOON

The inner core is a solid ball of pure iron only 240 km (150 miles) across

The outer core is a molten layer, consisting mainly of liquid iron

The lower mantle is a layer of partly molten rock, wrapped around the core

About a quarter the size of Earth, the Moon is its companion in space. Some astronomers refer to the Earth-Moon system as a "double planet". The Moon is airless, rugged, and covered in craters and jagged mountains. Because it is so close to Earth, the Moon's gravity causes tides in our seas.

LUNAR CRATERS

The Eratosthenes Crater is one of hundreds of craters that dot the Moon. They range in size from tiny pits to mountain-ringed dishes many hundreds of kilometres across. The Moon's surface has hardly changed in billions of years.

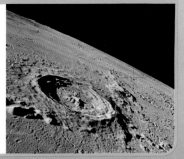

The outer mantle is made of iron-rich rock

Fragments from a **collision between Earth and a Mars-sized body** formed our Moon.

TIME TO ORBIT EARTH:
27.32 Earth days

SURFACE TEMPERATURE:
-247 to 120°C (-410 to 250°F)

The crust is dominated by a set of huge basins created by an asteroid bombardment 3.8 billion years ago

MARS

Mars, known as the Red Planet, is the world that most resembles our Earth. It is red, because it is, literally, rusty – from oxygen in the air reacting with iron minerals. Liquid water may still be present on Mars, leading to speculation that there is primitive life on the planet. Mars has been visited by several rovers, and orbiting spacecraft, and there are plans for humans to travel there by the 2030s.

Mars' surface is an endless desert of sand and rock

Giant volcano Olympus Mons is three times taller than Mount Everest

CANYON SYSTEM

Valles Marineris is a huge canyon system on Mars, created by volcanic activity in the Tharsis region that caused the planet's surface to crack apart. The gash is so vast that you could fit Earth's Alps inside it. It is 7 km (4 miles) deep and 4,000 km (2,480 miles) long: ten times longer than America's Grand Canyon.

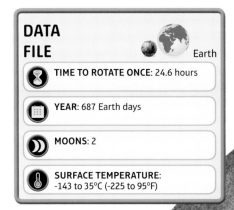

DATA FILE

Earth

TIME TO ROTATE ONCE: 24.6 hours

YEAR: 687 Earth days

MOONS: 2

SURFACE TEMPERATURE: -143 to 35°C (-225 to 95°F)

The mantle is a deep layer of silicate rock

Mars' small core is partially liquid, and made of iron

Mars' biggest volcano, **Olympus Mons is 22 km (13 miles) high**, and could almost cover France.

Mars' crust shows signs of ancient oceans and rivers

Valles Marineris

Mars' thin atmosphere is made of 95 per cent carbon dioxide

Our Solar System

THE ASTEROID BELT

Between the orbits of Mars and Jupiter lies a large region of rocky debris left over from the formation of our Solar System 4.6 billion years ago. Put together, these rocks, called asteroids, would weigh only one-thousandth of our planet's mass. Astronomers have discovered half a million of these asteroids, and spacecraft have taken close-up pictures of a dozen of them.

DATA FILE

DISTANCE FROM THE SUN: 270–620 million km (167–385 million miles)

YEAR OF DISCOVERY: 1801, when Giuseppe Piazzi discovered Ceres

SMALLEST ASTEROID: 145779 (1998 CC), 0.3 km (0.2 miles) in diameter

LARGEST ASTEROID: Ceres, 946 km (590 miles) in diameter

LARGEST ASTEROID

One asteroid, Ceres, is so much bigger than any other asteroid that some astronomers now call it a "dwarf planet". Its icy surface is heavily pitted by impacts, with a strange pyramid-shaped mountain 5 km (3 miles) high. Small white spots on Ceres might be patches of salt from an underground ocean.

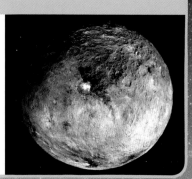

Asteroids are heavily cratered because of collisions

Some asteroids may be rich in valuable metals such as platinum

JUPITER

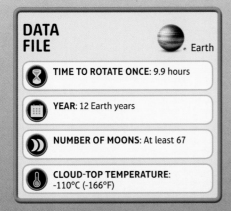

Jupiter is the king of the Solar System. This mighty planet could contain 1,300 Earths, and has a powerful magnetic field that causes spectacular light displays (called aurora) at its poles. Unlike solid, rocky planets, such as Earth, Jupiter is made mostly of gas. It has at least 67 moons. Its strong gravity has captured many asteroid-like moons.

Jupiter's colourful bands are made up of clouds at different heights

DATA FILE

Earth

⧗ **TIME TO ROTATE ONCE:** 9.9 hours

▦ **YEAR:** 12 Earth years

☽☽ **NUMBER OF MOONS:** At least 67

🌡 **CLOUD-TOP TEMPERATURE:** -110°C (-166°F)

The Great Red Spot is an enormous storm that has raged for more than two centuries

Another layer of liquid hydrogen lies below Jupiter's clouds

GREAT RED SPOT

Jupiter's most famous feature, the Great Red Spot is a swirling storm that was once three times the size of Earth, although it has recently started shrinking. Its red colour is probably caused by sunlight breaking up chemicals in the tops of the highest clouds.

Above the core is a layer of hydrogen so compressed that it behaves like a liquid metal

Jupiter almost certainly has a solid core, around which the rest of the planet formed

The atmosphere is mainly hydrogen and some helium, with a cocktail of methane, ammonia, water, and hydrogen sulfide

Despite its size, **Jupiter has the shortest day** in the Solar System.

SATURN

Saturn's spectacular rings are so wide they would stretch almost from Earth to the Moon. They are an amazing sight through a telescope. Made of ice-covered debris with particles ranging in size from ice cubes to houses, the rings are probably the remains of a moon torn apart by Saturn's gravity.

Saturn's atmosphere is mostly hydrogen and helium, with clouds of ammonia ice and water ice on top

Saturn's rings orbit the planet in an almost perfectly flat plane

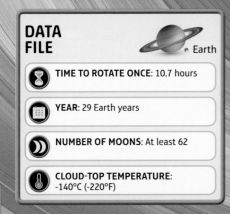

DATA FILE

Earth

⏳ **TIME TO ROTATE ONCE**: 10.7 hours

📅 **YEAR**: 29 Earth years

🌙 **NUMBER OF MOONS**: At least 62

🌡️ **CLOUD-TOP TEMPERATURE**: -140°C (-220°F)

Liquid hydrogen
lies below
Saturn's clouds

MIGHTY TITAN

Saturn has at least 62 moons – the biggest being Titan, which is larger than Mercury. In 2005, the Huygens probe penetrated the moon's thick clouds and landed on its frozen surface. Titan has huge lakes of ethane and methane.

The gas giant's core may be solid or liquid, or a mix of both

Compressed hydrogen, behaving like liquid metal, surrounds the core

Saturn's **density is so low** that, if it were put in an ocean, it would float.

URANUS

Astronomer William Herschel discovered Uranus in 1781. Four times wider than Earth, this gas giant is circled by 13 thin rings. When the Voyager 2 probe passed the planet in 1986, it gazed upon a bland, blue world, toppled on its side. Uranus's odd orientation causes extremes in sunlight reaching the planet, creating 20-year-long seasons. As the atmosphere heated up, giant storms appeared in its clouds in August 2014.

The atmosphere on Uranus is mainly hydrogen and helium

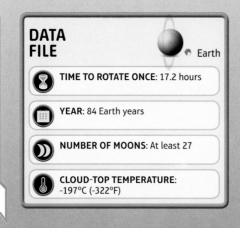

DATA FILE

Earth

TIME TO ROTATE ONCE: 17.2 hours

YEAR: 84 Earth years

NUMBER OF MOONS: At least 27

CLOUD-TOP TEMPERATURE: -197°C (-322°F)

Unlike the dazzling rings of Saturn, Uranus's 13 rings are narrow, and as black as coal

URANUS'S MOONS

Uranus boasts at least 27 moons. The most remarkable is tiny Miranda, just 500 km (310 mile) across. Its crumpled surface is riddled with grooves, craters, and cliffs. The moon was probably smashed apart by a huge impact in the past, before it re-formed. Uranus hangs in Miranda's sky in this artist's impression.

At 5,000°C (9,000°F), Uranus's core is nearly as hot as the Sun's surface

The planet's mantle is a slushy mixture of water, ammonia, and methane

Uranus **rolls on its side** as it orbits the Sun.

x

NEPTUNE

Neptune was the last planet to be discovered, and was found thanks to the power of maths. Astronomers had discovered that Uranus wandered off its predicted path as though pulled by an unknown planet's gravity. When they calculated where the mystery planet should be and looked through a telescope, they found Neptune. This gas-giant is noticeably active. It has a hot core that drives its storm-systems and dark spots.

The planet's atmosphere is made of hydrogen and helium, with methane providing the blue colour

DATA FILE

Earth

TIME TO ROTATE ONCE: 16.1 hours

YEAR: 165 Earth years

NUMBER OF MOONS: At least 14

CLOUD-TOP TEMPERATURE: -201°C (-330°F)

Neptune's hot core may be surrounded by an ocean of liquid diamonds

Most of the planet's mass lies in the mantle, which is a layer composed of a deep ocean of water, ammonia, and methane

ICE-COLD TRITON

At 2,700 km (1,680 miles) across, Triton is Neptune's biggest moon. It is one of the coldest places in the Solar System. It has geysers that erupt plumes of nitrogen and dust into space. Triton orbits in the "wrong" direction, moving the opposite way to Neptune's rotation. This suggests Triton was captured by the planet's gravity.

One of Neptune's five faint rings, made mostly of dust

Winds on Neptune reach over 2,000 kph (1,240 mph), the **fastest in the Solar System**.

The Great Dark Spot, a dramatic cloud feature, spotted by Voyager 2 when it swung by the planet in 1989, disappeared quickly, like most of Neptune's weather features

THE KUIPER BELT

The Kuiper Belt is a vast zone beyond the orbit of Neptune with large icy bodies. When Pluto was discovered, astronomers called it "planet nine", but we now know it is just one of the thousands of Kuiper Belt Objects (KBOs). In 2015, the New Horizons spacecraft showed Pluto has icy mountains and plains of frozen nitrogen. The craft is now heading to a KBO called 2014MU$_{69}$.

DATA FILE

DISTANCE FROM SUN:
4.5–10 billion km (2.8–6 billion miles)

FIRST KBO DISCOVERED: Pluto (1930)

SMALLEST KBO: 65407 (2002 RP120), 15 km (9 miles) in diameter

LARGEST KBO: Pluto, 2,370 km (1,470 miles) in diameter

Largely made of ice, mixed with rocky pebbles, the temperature on a KBO is usually -220°C (-364°F)

DWARF PLANETS

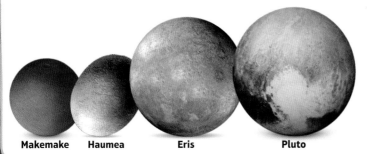

Makemake **Haumea** **Eris** **Pluto**

The four largest Kuiper Belt Objects are called "dwarf planets", because, like the planets, their gravity is powerful enough to pull them into a spherical shape. Pluto is the biggest, slightly smaller Eris is the heaviest, while the others are Haumea and Makemake.

Pluto's name was suggested by an **11-year-old British schoolgirl**.

Fresh white ice – made of frozen water, methane, and ammonia – may be exposed after a collision

The surface of a KBO turns reddish after exposure to the radiation in space

This giant belt of icy bodies extends 30–50 times further from the Sun than our planet. A larger icy zone called the Oort Cloud lies beyond this belt.

COMETS

With its great glowing tail, a brilliant comet is an awesome sight. Ancient civilizations saw these streaks of light in the sky as omens of doom, but a comet is simply a large chunk of ice and dust orbiting the Sun. Comets come from the edge of the Solar System – in the Oort Cloud, which is home to billions of comets.

In 1996, Comet Hyakutake made one of the closest approaches of any comet in the last 200 years. Seen by many people, it was easily visible in the night sky.

The small, solid nucleus erupts into life as it nears the Sun

Pressure from the Sun's light pushes dust from the comet out into a tail that reflects the sunlight.

Gas and dust from the nucleus forms a "head" (coma) that can grow as large as the Sun

INSIDE A COMET

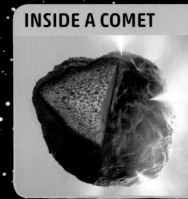

The heart of a comet is its solid nucleus, shown in this cut-away image. This "dirty snowball", typically a few kilometres across, is made of ice and rocky dust left over from the birth of the Solar System. When heated by the Sun's warmth, jets of gas burst through the dark, rocky crust.

Particles streaming outwards from the Sun propel the comet's gas into a second, narrower tail

A chunk from a comet hit Earth 66 million years ago, **wiping out the dinosaurs**.

DATA FILE

 MAXIMUM DISTANCE FROM THE SUN: 10 trillion km (6.2 trillion miles)

 **SMALLEST NUCLEUS**: Comet 322P/ SOHO, 0.2 km (0.12 miles) in diameter

LARGEST NUCLEUS: Comet Hale-Bopp, 60 km (37 miles) in diameter

EXOPLANETS

Planets outside our Solar System are called exoplanets. To date, astronomers have found more than 3,000 exoplanets orbiting other stars beyond our Solar System. So far, these other planets are mainly much bigger or much hotter than our Earth. But the search is on for a planet similar to Earth – a warm, wet world, that may be home to some form of alien life.

HD 189733 A

Astronomers think there are **40 billion Earth-like planets** in the Milky Way.

The star's planet, HD 189733 b, is similar in size to Jupiter, but is roasted by its proximity to its parent star

THE SEARCH FOR ALIEN LIFE

The dishes of the Allen Telescope Array in California, USA, form a highly sensitive "ear" on the Universe, listening out for a radio broadcast from any intelligent aliens that may live on distant planets. The Array is part of an international project called SETI (Search for Extraterrestial Intelligence).

SMALLEST EXOPLANET: Kepler-37b, 3,860 km (2,400 miles) in diameter

LARGEST EXOPLANET: HAT-P-32b, 291,000 km (180,800 miles) in diameter

COLDEST EXOPLANET: OGLE-2005-BLG-390Lb, -220°C (-364°F)

HOTTEST EXOPLANET: Kepler-70b, 6,900°C (12,450°F)

Slightly smaller than our Sun, the orange dwarf HD 189733 A lies 63 light years away from Earth. It is only one-third as bright as the Sun.

Our Solar System

3

DISCOVERING THE UNIVERSE

Throughout history, people have endeavoured to understand the mysteries of the Universe, first with just their eyes and minds, and then with increasingly powerful telescopes. Astronomers have tracked down evidence of the Big Bang, and discovered giant galaxies, deadly black holes, newborn and exploding stars, and thousands of distant planets, some of which may be home to alien life.

PLATO

428–348 BCE

Foremost among the Greek philosopher-scientists, Plato founded an Academy, the equivalent to a top university, in Athens. Students studied philosophy and science, and worked out in the gymnasium. Plato had a particular interest in the movement of the planets. He could not figure out why some of them moved in loops as they orbited Earth.

A sign over Plato's Academy read, **"Let no one ignorant of geometry enter here."**

PTOLEMY

c.85–168 CE

Celestial map based on constellations named by Ptolemy

Ptolemy also wrote *Harmonics*, a work **about the theory of music**.

Ptolemy was a philosopher-scientist living in Egypt, at that time part of the Greek Empire that stretched from Greece to India. He was responsible for collating *The Almagest*, a collection of Greek scientific knowledge, that would stand the test of time for 1,500 years. He also listed 1,022 stars, and 48 constellations still in use today.

DATA FILE

KEY ACHIEVEMENT: Published *The Almagest*, a scientific text

COUNTRY OF ORIGIN: Egypt

YEAR OF ACHIEVEMENT: 150 CE

Discoverers of the Universe

OMAR KHAYYAM

1048-1131

After the fall of the Greek Empire, Persia became the major centre for science and philosophy. Among the most influential thinkers was Omar Khayyam. His study of astronomy in 1079 led to him devising a calendar that is just as accurate as the one used in the West today. It is still used in some Islamic countries, 1,000 years later.

Khayyam **was also a poet**. His verses have been translated into many languages.

DATA FILE

KEY ACHIEVEMENT: Devised an accurate calendar with a year lasting 365.24 days

COUNTRY OF ORIGIN: Persia (modern-day Iran)

YEAR OF ACHIEVEMENT: 1079

NICOLAUS COPERNICUS

1473–1543

DATA FILE

🏆 **KEY ACHIEVEMENT:** Discovered that Earth orbited the Sun

🚩 **COUNTRY OF ORIGIN:** Poland

📅 **YEAR OF ACHIEVEMENT:** 1543

Copernican model showing the planets orbiting the Sun

Copernicus **was a canon at Frombork Cathedral**, in modern-day Poland.

By the 16th century, Europe had become the centre of scientific knowledge. In 1543, Copernicus concluded that the planets circled the Sun, not Earth. This was a revolutionary and dangerous idea that contradicted the Church's view that Earth was stationary at the centre of the Universe.

Discoverers of the Universe

JOHANNES KEPLER

1571–1630

A great mathematician, Kepler calculated that each planet orbits the Sun in an oval path called an ellipse, travelling fastest when it is closest to the Sun, and slowest when it is further away. Kepler also discovered that the motions of planets depended on their distance from the Sun, but he did not know the reason why.

Kepler believed Earth had a soul, which affected weather and other phenomena.

Kepler's model of planetary orbits defended Copernicus's theory

DATA FILE

KEY DISCOVERY:
Laws of planetary motion

COUNTRY OF ORIGIN: Germany

YEAR OF DISCOVERY: 1609

GALILEO GALILEI

1564–1642

Ganymede, a moon of Jupiter discovered by Galileo

Galileo's **shrivelled fingers** are displayed at the Science Museum in Florence, Italy.

Galileo's improvement of the design of the telescope meant parts of the Solar System could be seen clearly for the first time. He observed mountains on the Moon and the four biggest moons of Jupiter. He also found Venus to have phases like our Moon. This led him to believe that the Sun is at the centre of the Solar System. The Church put Galileo on trial for his radical views.

DATA FILE

KEY ACHIEVEMENT: Proved Earth orbited the Sun

COUNTRY OF ORIGIN: Italy

YEAR OF ACHIEVEMENT: 1610

CHRISTIAAN HUYGENS
1629-1695

Huygens also **invented** the **pendulum clock** and the **pocket watch**.

In 1659, Huygens published the idea that Saturn was surrounded by a flat ring

Dutch astronomer Christiaan Huygens developed lenses that made telescopes more powerful. In 1655, he spotted dark markings on Mars, and discovered the planet was spinning. He also discovered the rings around the planet Saturn, but he did not know what they were made of.

DATA FILE

🏆 **KEY DISCOVERY**: Established that Saturn was surrounded by rings

🚩 **COUNTRY OF ORIGIN:** The Netherlands

▦ **YEAR OF DISCOVERY**: 1655

ISAAC NEWTON

1642-1726

In 1687, Isaac Newton became the first scientist to understand that the movement of the planets was controlled by gravity, a force that attracts all objects that have mass. The closer the bodies, the stronger the force. Even now, physicists do not understand how gravity works.

Newton was also **Master of the Royal Mint**, preventing the forgery of coins.

DATA FILE

KEY ACHIEVEMENT: The first to understand gravity

COUNTRY OF ORIGIN: UK

YEAR OF DISCOVERY: 1687

Replica of Isaac Newton's telescope

EDMOND HALLEY

1656–1742

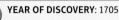
Halley invented the diving bell, in which he plunged 18.3 m (60 ft) below the River Thames.

Edmond Halley devoted only a portion of his time to astronomy. He was also a mathematics professor, a sea captain, and a spy. But his name is forever linked to the comet named after him. After studying its orbit around the Sun, he predicted that it would return in 1758, 53 years later. It did and every 75 to 76 years since.

Halley's Comet

Discoverers of the Universe

WILLIAM HERSCHEL

1738–1822

Astronomer William Herschel's discovery of Uranus in 1781, and two of its moons in 1787, made with a homemade telescope, increased the number of known planets from six to seven. In doing so, he doubled the size of our Solar System – Uranus was nearly twice as far from the Sun as Saturn, the sixth planet from the Sun. Appointed as the King's Astronomer by King George III, he also studied the structure of the Milky Way.

Herschel's sister, Caroline, was an astronomer, too. She **discovered eight comets**.

The Great 40-foot telescope, built and used by Herschel

DATA FILE

KEY DISCOVERIES: Identified Uranus and its moons Titania and Oberon

COUNTRY OF ORIGIN: Germany

YEARS OF DISCOVERIES: 1781 (Uranus) and 1787 (its moons)

KONSTANTIN TSIOLKOVSKY

1857–1935

Tsiolkovsky, **the most prominent crater on the Moon's far-side**, is named after him.

Konstantin Tsiolkovsky is regarded as the father of spaceflight. He was fascinated by the concept and design of rockets, but was not taken seriously by his fellow scientists. Though he never launched a rocket, he imagined the possibilities of human spaceflight. In 1903, he published a book – his key work on rocketry – and its ideas would inspire and contribute towards the work of future generations of rocket scientists.

DATA FILE

KEY ACHIEVEMENT: Developed theory of spaceflight

COUNTRY OF ORIGIN: Russia

YEAR OF ACHIEVEMENT: 1903

EDWIN HUBBLE

1889–1953

Hubble was an **amateur boxer, and trained as a tank driver.**

In 1929, astronomer Edwin Hubble discovered that other galaxies are moving away from us. He concluded that the Universe was expanding from a huge, hot explosion, now known as the "Big Bang". By studying the speeds at which these galaxies were moving, astronomers calculated that our Universe was born almost 14 billion years ago.

DATA FILE

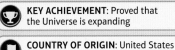

KEY ACHIEVEMENT: Proved that the Universe is expanding

COUNTRY OF ORIGIN: United States

YEAR OF ACHIEVEMENT: 1929

Discoverers of the Universe

CECILIA PAYNE-GAPOSCHKIN
1900–1979

Cecilia Payne-Gaposchkin was a brilliant female astronomer. In 1925, after analysing the light from stars using an instrument called a spectrometer, she started to understand what stars were made of. She discovered that they largely consisted of hydrogen, now known to be the most common element in the Universe.

Cecilia **studied at Cambridge in the 1920s** – a time when women were not awarded degrees.

DATA FILE

KEY DISCOVERY: Established that stars are mainly made of hydrogen

COUNTRY OF ORIGIN: UK

YEAR OF ACHIEVEMENT: 1925

FRED HOYLE

1915–2001

A prolific science-fiction writer, **Hoyle wrote the hit TV show** *A for Andromeda*.

In 1957, Fred Hoyle made a major breakthrough when he worked out how elements form in the centre of a star. He was a maverick among astronomers. Hoyle never agreed with what was commonly believed, and dismissed the "Big Bang" theory, believing instead in a "steady-state" Universe that was unchanging, although that theory has been proved to be wrong.

DATA FILE

- **KEY ACHIEVEMENT:** Explained how elements are produced in stars
- **COUNTRY OF ORIGIN:** UK
- **YEAR OF ACHIEVEMENT:** 1957

LAIKA THE DOG

1954-1957

Laika, a little dog who orbited Earth in the second satellite to be launched, blazed the trail for humans to fly into space. Unfortunately, she did not survive the journey on this Russian mission, as her craft's cooling system failed. The next canine crew of Belka and Strelka landed safely and even had puppies.

Laika was a stray dog, **found roaming the streets** of Moscow.

DATA FILE

KEY ACHIEVEMENT: First animal to orbit Earth

COUNTRY OF ORIGIN: Russia

YEAR OF ACHIEVEMENT: 1957

YURI GAGARIN

1934–1968

Gagarin's short height was **essential** to fit into the capsule.

History was made on 12 April 1961 when, for the first time, a human left our planet and travelled into space. Yuri Gagarin, a 27-year-old Russian pilot, was launched into space in a capsule called Vostok 1. Gagarin made one orbit of Earth in 108 minutes and landed safely.

DATA FILE

🏆 **KEY ACHIEVEMENT:** First human in space

🚩 **COUNTRY OF ORIGIN:** Russia

📅 **YEAR OF ACHIEVEMENT:** 1961

JOHN GLENN

Born 1921

In 1998, at the age of 77, Glenn flew on a space shuttle to become **the oldest-ever astronaut**.

DATA FILE

🏆 **KEY ACHIEVEMENT**: First American to orbit Earth

🚩 **COUNTRY OF ORIGIN**: United States

📅 **YEAR OF ACHIEVEMENT**: 1962

John Glenn made three orbits of our planet in 1962 to become the first American to orbit Earth. As he passed over Australia in the dark, the residents of Perth switched on all their lights to greet him. Glenn's capsule, Friendship 7, splashed down safely in the Atlantic Ocean and he was picked up by a navy ship.

VALENTINA TERESHKOVA

Born 1937

19 years would pass before another **female astronaut flew in space**.

An amateur skydiver, Valentina Tereshkova spent three days in space in 1963, completing 48 orbits of Earth. Part of her mission was to fly her spacecraft in formation with another Russian capsule. She later became a politician and, in 2013, volunteered to go on a one-way journey to Mars.

DATA FILE

KEY ACHIEVEMENT: First woman in space

COUNTRY OF ORIGIN: Russia

YEAR OF ACHIEVEMENT: 1963

MAARTEN SCHMIDT

Born 1929

Artist's impression of a quasar

"Quasar" is short for **"quasi-stellar radio source"**.

Astronomers had identified mysterious, star-like objects that emitted radio waves. In 1963, Maarten Schmidt found that one of these, 3C 273, lies billions of light years away. For it to be visible, it must be extremely bright. Astronomers called objects like this "quasars", short for "quasi-stellar radio source".

DATA FILE

KEY DISCOVERY: Discovered quasars

COUNTRY OF ORIGIN: The Netherlands

YEAR OF ACHIEVEMENT: 1963

ALEXEI LEONOV

Born 1934

DATA FILE

🏆 **KEY ACHIEVEMENT**: First human to carry out a spacewalk

🚩 **COUNTRY OF ORIGIN**: Russia

📅 **YEAR OF ACHIEVEMENT**: 1965

A skilled artist, **Leonov pencil-sketched** the views from space.

In 1965, Russian cosmonaut Alexei Leonov carried out the first-ever "spacewalk" or EVA (extravehicular activity). He was outside the spacecraft Voskhod 2 for 12 minutes and 9 seconds. In the vacuum of space his suit had inflated and he was unable to pass back into the airlock until he opened a valve to release the pressure.

Artist's impression of Leonov's first spacewalk

ARNO PENZIAS AND ROBERT WILSON

Born 1933 and 1936

In 1964, using a giant antenna in New Jersey, United States, Arno Penzias and Robert Wilson detected a faint noise coming from all over the sky. They realized that the noise was radiation – leftover heat from a giant explosion, emitted by hot gas that filled the early Universe. Named "Cosmic Microwave Background", the radiation was clinching evidence for the Big Bang.

Arno Penzias

Robert Wilson

Their discovery won the pair a **Nobel Prize in 1978**.

DATA FILE

 KEY DISCOVERY: Proved the Big Bang theory

COUNTRY OF ORIGIN: United States

YEAR OF DISCOVERY: 1964

JOCELYN BELL BURNELL AND ANTONY HEWISH

Born 1943 and 1924

In 1967, using the radio telescope designed by Antony Hewish, Jocelyn Bell Burnell discovered regular "pulses" of radio waves coming from unknown objects in space. These compact neutron stars, that spin so rapidly they appear to flash like a lighthouse, were named "pulsars".

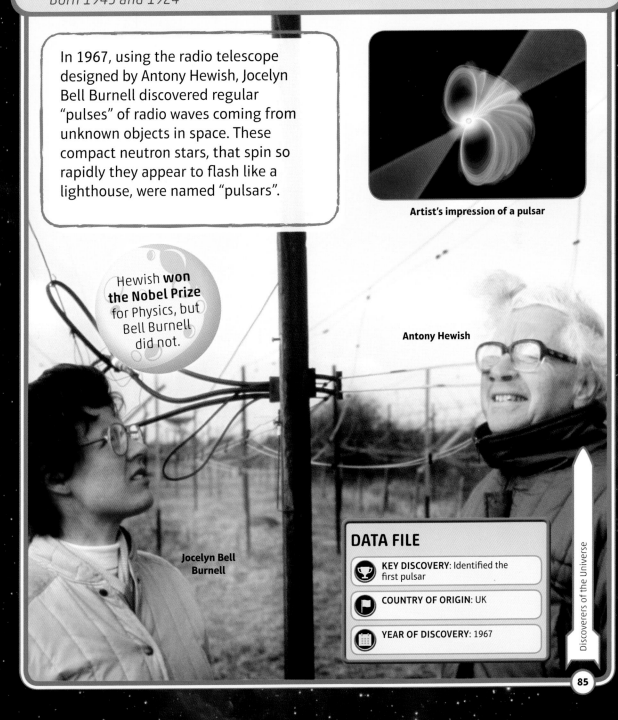

Artist's impression of a pulsar

Hewish **won the Nobel Prize** for Physics, but Bell Burnell did not.

Antony Hewish

Jocelyn Bell Burnell

DATA FILE

🏆 **KEY DISCOVERY:** Identified the first pulsar

🚩 **COUNTRY OF ORIGIN:** UK

📅 **YEAR OF DISCOVERY:** 1967

NEIL ARMSTRONG

1930–2012

A former test-pilot renowned for his calmness under pressure, Neil Armstrong commanded Apollo 11, the first manned mission to the Moon, in 1969. After a difficult landing, Armstrong then stepped onto the lunar surface saying, "That's one small step for man, one giant leap for mankind."

A decade before becoming an astronaut, Armstrong **flew 78 combat missions** in the Korean War.

DATA FILE

KEY ACHIEVEMENT: First man to walk on the Moon

COUNTRY OF ORIGIN: United States

YEAR OF ACHIEVEMENT: 1969

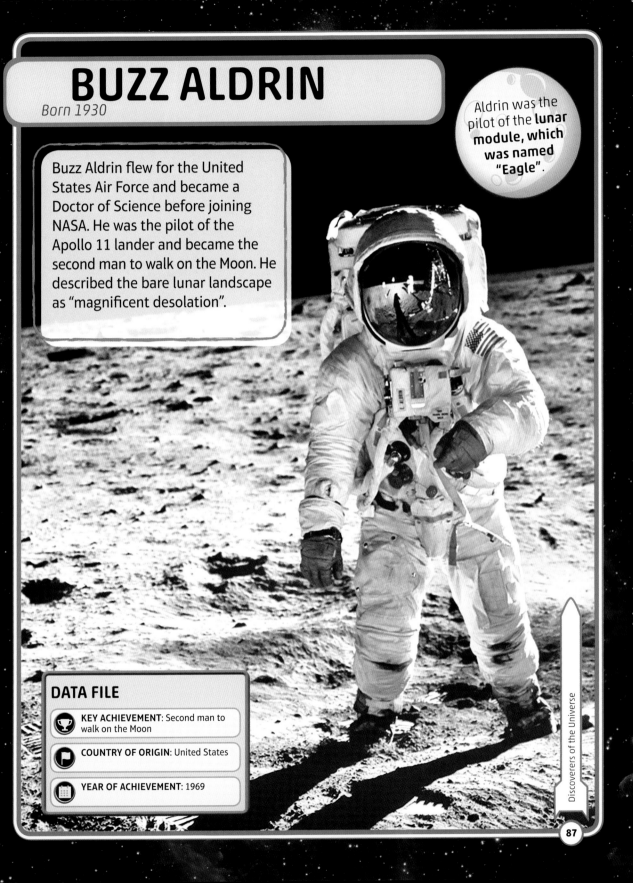

BUZZ ALDRIN
Born 1930

Aldrin was the pilot of the **lunar module, which was named "Eagle"**.

Buzz Aldrin flew for the United States Air Force and became a Doctor of Science before joining NASA. He was the pilot of the Apollo 11 lander and became the second man to walk on the Moon. He described the bare lunar landscape as "magnificent desolation".

DATA FILE

KEY ACHIEVEMENT: Second man to walk on the Moon

COUNTRY OF ORIGIN: United States

YEAR OF ACHIEVEMENT: 1969

Discoverers of the Universe

VALERI POLYAKOV

Born 1942

An expert in the field of space medicine, Valeri Polyakov spent long periods in Earth's orbit on the Russian space station Mir. After a first mission of eight months from August 1988 to April 1989, he returned to Mir nearly five years later and spent 14 months on the space station. The 437.7 days he spent on the second trip is a record.

During his time in space, Polyakov travelled **300 million km** (187 million miles) in Mir.

Mir Space Station

DATA FILE

KEY ACHIEVEMENT: Longest time continuously spent in space

COUNTRY OF ORIGIN: Russia

YEAR OF ACHIEVEMENT: 1995

MICHEL MAYOR AND DIDIER QUELOZ
Born 1942 and 1966

When, in 1995, astronomers Michel Mayor and Didier Queloz were observing the star 51 Pegasi, they found out that it was being pulled back and forth by a planet. This planet (51 Pegasi b) – the first to be found around a Sun-like star, outside our Solar System – is almost as heavy as Jupiter, but it is so close to its parent star that it has been baked to almost 1,000°C (1,830°F). As a result, it is called a "hot Jupiter".

Hot Jupiters orbit about 100 times closer to their parent star than Jupiter does to the Sun.

Didier Queloz

Michel Mayor

DATA FILE

KEY DISCOVERY: Found first planet around a Sun-like star

COUNTRY OF ORIGIN: Switzerland

YEAR OF DISCOVERY: 1995

BRIAN SCHMIDT AND SAUL PERLMUTTER

Born 1967 and 1959

Brian Schmidt

Einstein predicted dark energy, but then dismissed it as a blunder.

Saul Perlmutter

Brian Schmidt and Saul Perlmutter led rival teams that were trying to measure how distant galaxies in the expanding Universe were slowing down under the influence of gravity. In 1998, to their amazement, they found that the expansion of the Universe is actually speeding up. It has since been worked out that this is being caused by a mysterious force called "dark energy".

DATA FILE

KEY DISCOVERY: The expansion of the Universe is speeding up

COUNTRY OF ORIGIN: United States

YEAR OF DISCOVERY: 1998

DENNIS TITO
Born 1940

Tito agreed to **pay for any breakages** he caused on board the ISS.

Although he became a multi-millionaire through his success in business, Dennis Tito always loved space. He had previously designed an unmanned Mars-mission for NASA. Tito paid $20 million to travel with a Russian crew on a mission to the International Space Station (ISS) for a week, and, in 2001, became the first person to buy his own ticket to space.

DENNIS A. TITO
ДЕННИС А. ТИТО

DATA FILE

🏆 **KEY ACHIEVEMENT:** First space tourist

🏳 **COUNTRY OF ORIGIN:** United States

📅 **YEAR OF ACHIEVEMENT:** 2001

YANG LIWEI

Born 1965

The Chinese word for astronaut is **"taikonaut"**.

A former fighter pilot, Yang became the first man from China in space in 2003 when he flew the Shenzhou 5 capsule for 14 orbits around Earth. Until then, only Russia and the United States had undertaken manned space missions. Yang's flight marked the start of China's manned space programme, and the country has plans for a space station in the 2020s.

Artist's impression of the Shenzhou 5 craft

DATA FILE

KEY ACHIEVEMENT: The first Chinese astronaut to go into space

COUNTRY OF ORIGIN: People's Republic of China

YEAR OF ACHIEVEMENT: 2003

KIP THORNE AND RAINER WEISS

Born 1940 and 1932

Kip Thorne predicted the existence of wormholes, and also calculated that black holes can produce ripples in space – just as a stone dropped in a pond makes ripples in the water. Rainer Weiss devised the Laser Interferometer Gravitational-Wave Observatory (LIGO), which, in 2016, picked up the ripples from two colliding black holes.

DATA FILE

KEY DISCOVERY: Identified and observed gravitational waves

COUNTRY OF ORIGIN: United States

YEAR OF DISCOVERY: 2016

LIGO can measure ripples **one-billionth** the size of an atom.

Rainer Weiss

Kip Thorne

Artist's impression of gravitational waves

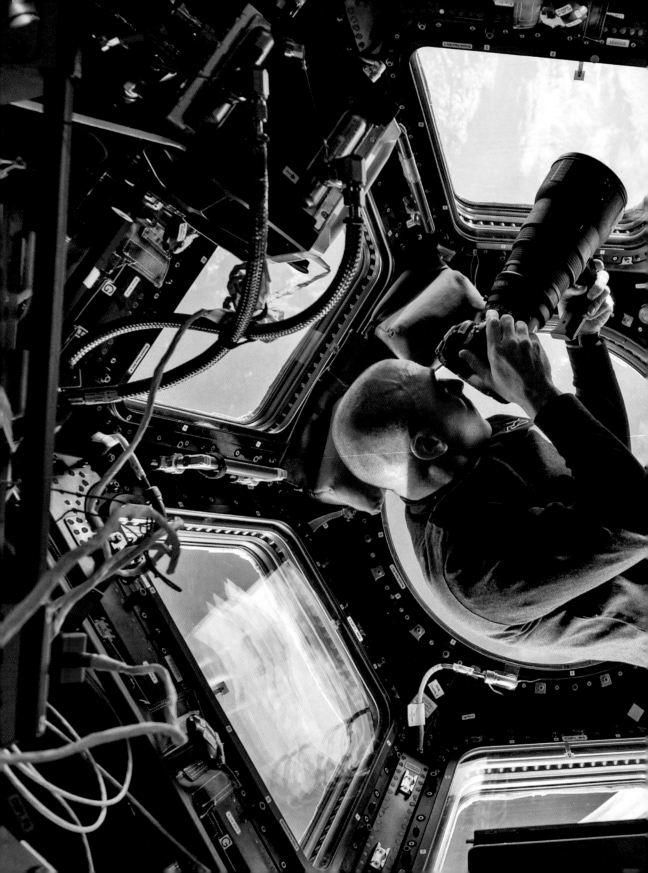

4

SPACE EXPLORATION

Exploring space is perhaps the greatest adventure of all. In this great age of space exploration, hardy robotic probes have voyaged to comets, asteroids, moons, and planets, while intrepid human explorers have not only orbited our planet, but have also walked on the Moon, lived and worked for long periods on space stations, and may be only decades away from flying to Mars and beyond.

V-2 ROCKET

Rocket

The Germans developed the V-2 rocket as a military missile during World War II. After Germany's defeat, a number of V-2s were captured and taken to the United States, along with their principal designer, the brilliant Wernher von Braun. The V-2s paved the way for America's civilian space programme.

The V-2 rocket was the first man-made object to **pass the boundary of space**.

DATA FILE

SIZE: 14 m (45 ft) long; 1.65 m (5.5 ft) in diameter

WEIGHT: 12,500 kg (27,600 lb)

FIRST LAUNCHED: 12 August 1943

COUNTRY: Germany and, later, United States

DANGER
GEVAAR

DANGER
V 2
KEEP CLEAR

Military display of captured German V-2 rocket in London after World War II

SPUTNIK 1
Satellite

Launched by Russia on 4 October 1957, Sputnik 1 was the first man-made object to orbit Earth. The size of a beach ball, this metal satellite circled the world for more than two months before burning up when it re-entered our planet's atmosphere. Sputnik 1's launch triggered a "space race" between Russia and the United States.

DATA FILE

SIZE: 58 cm (23 in) in diameter

DATE OF LAUNCH: 4 October 1957

MISSION DURATION: 92 days (de-orbited)

SPACE AGENCY: Russian space program

The 2.9 m (9.5 ft) antennae sent the first signals from orbit down to Earth

The United States launched their first satellite, **Explorer, four months later**.

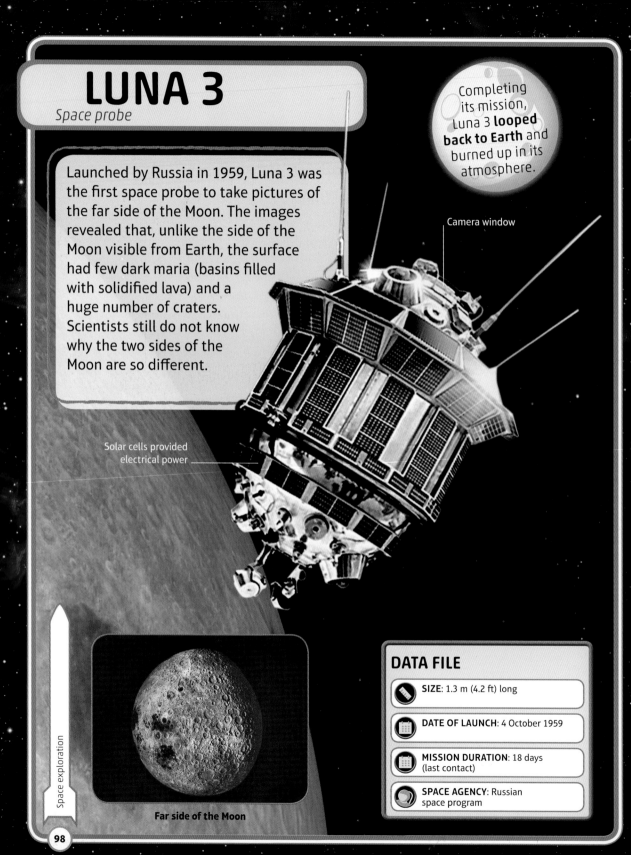

LUNA 3
Space probe

Completing its mission, Luna 3 **looped back to Earth** and burned up in its atmosphere.

Launched by Russia in 1959, Luna 3 was the first space probe to take pictures of the far side of the Moon. The images revealed that, unlike the side of the Moon visible from Earth, the surface had few dark maria (basins filled with solidified lava) and a huge number of craters. Scientists still do not know why the two sides of the Moon are so different.

Camera window

Solar cells provided electrical power

Far side of the Moon

DATA FILE

SIZE: 1.3 m (4.2 ft) long

DATE OF LAUNCH: 4 October 1959

MISSION DURATION: 18 days (last contact)

SPACE AGENCY: Russian space program

VOSTOK 1
Manned spacecraft

Yuri Gagarin

Vostok 1 was the first craft to carry a human being – Russian cosmonaut Yuri Gagarin – into space. On 12 April 1961, Gagarin spent over an hour orbiting Earth. He landed by parachute in the Saratov region of Russia, 280 km (174 miles) from the intended site at Baikonur – from where Vostok 1 had been launched.

The entire flight was controlled from the **Russian space centre** on Earth.

Spherical tanks held oxygen and nitrogen for life support

DATA FILE

SIZE: 2.3 m (7.5 ft) in diameter

DATE OF LAUNCH: 12 April 1961

MISSION DURATION: 1 hour, 48 minutes

SPACE AGENCY: Russian space program

MARINER 4
Space probe

Launched by NASA in November 1964, Mariner 4 was the first space probe to make a successful flyby of Mars. It also sent back the first close-up images of the Red Planet. Scientists were hopeful of seeing signs of life – even vegetation – but all the space probe revealed was a dead, cratered world, much like our Moon.

DATA FILE

SIZE: 2.89 m (9.5 ft) long

DATE OF LAUNCH: 28 November 1964

MISSION DURATION: 3 years, 23 days (deactivated)

SPACE AGENCY: NASA

It took Mariner 4 **seven-and-a-half months** to reach Mars.

A small rocket ensured Mariner 4 stayed on course

Surface of Mars as seen by Mariner 4

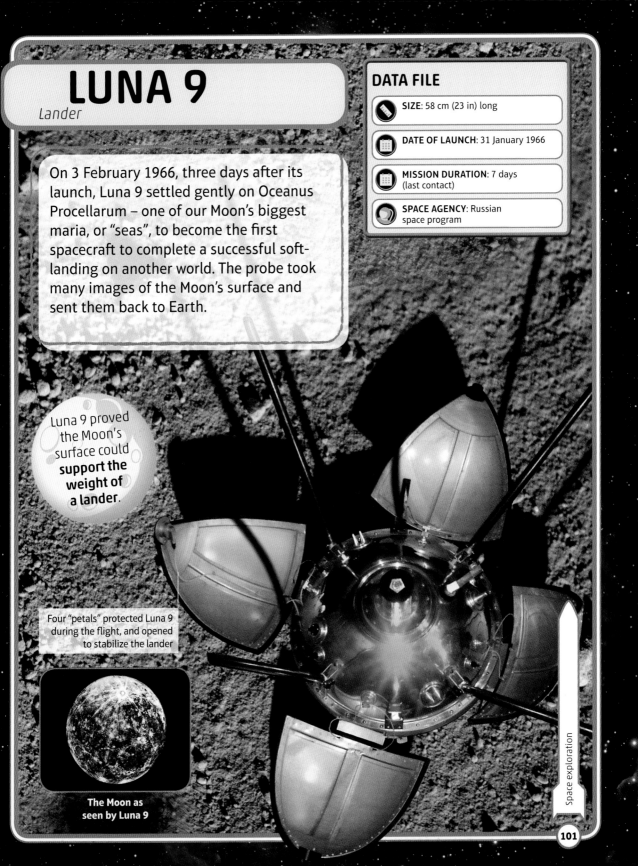

LUNA 9
Lander

DATA FILE

SIZE: 58 cm (23 in) long

DATE OF LAUNCH: 31 January 1966

MISSION DURATION: 7 days (last contact)

SPACE AGENCY: Russian space program

On 3 February 1966, three days after its launch, Luna 9 settled gently on Oceanus Procellarum – one of our Moon's biggest maria, or "seas", to become the first spacecraft to complete a successful soft-landing on another world. The probe took many images of the Moon's surface and sent them back to Earth.

Luna 9 proved the Moon's surface could **support the weight of a lander**.

Four "petals" protected Luna 9 during the flight, and opened to stabilize the lander

The Moon as seen by Luna 9

SATURN V
Rocket

The biggest and heaviest launch vehicle ever built, Saturn V was the rocket that took humans to the Moon. Built in the United States by a team led by Wernher von Braun – the great German rocket designer – it weighed 2,970 tonnes, and could blast a crew of three astronauts into space.

Escape rocket pulls command module to safety in event of faulty launch

Command module carried three astronauts

This section housed the lunar module, which landed on the Moon's surface

The command module was the **only part of the Apollo craft to return from the Moon**.

DATA FILE

SIZE: 111 m (364 ft) tall

FIRST LAUNCH: 9 November 1967 (Apollo 4 mission)

NUMBER OF MISSIONS: 13 (between 1967 and 1973)

SPACE AGENCY: NASA

APOLLO 8
Manned lunar orbiter

DATA FILE

SIZE: Command/service module – 11 m (36 ft) long

DATE OF LAUNCH: 21 December 1968

MISSION DURATION: 6 days, 3 hours, 40 minutes

SPACE AGENCY: NASA

NASA's Apollo 8 mission was a dress-rehearsal for landing humans on the Moon. Its crew comprised three astronauts: Frank Borman, James Lovell, and William Anders. The team were the first humans to leave Earth's orbit, and spent three days circling the Moon. They saw the far side of the Moon and took 700 images of its surface.

The Apollo 8 crew were the **first to see an "Earthrise"** above the Moon.

The service module contained oxygen and hydrogen to generate electricity, along with fuel

The rocket engine was used to place the spacecraft into lunar orbit, and, later, to propel it back towards Earth

Apollo 8 crew (left to right): James Lovell, William Anders, and Frank Borman

Space exploration

APOLLO 11
Manned lunar landing

On 16 July 1969, Apollo 11, the mission to land the first humans on the Moon, blasted off. Three days later, Neil Armstrong became the first man to set foot on the lunar surface. Buzz Aldrin joined him 20 minutes later, while Michael Collins remained in orbit in the Apollo command module. Armstrong and Aldrin spent 21 hours, 36 minutes on the Moon's surface.

Buzz Aldrin

The Passive Seismic Experiment detected moonquakes

The Apollo 11 crew **returned with 21.55 kg (47.5 lb) of moon rock**.

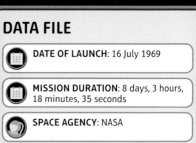

The lower section of the 7 m (23 ft) tall Lunar Module was left behind on the Moon, while the astronauts blasted off in the top section

ONE SMALL STEP

The first human footprints on the Moon could last forever, as the Moon has no atmosphere and, therefore, no wind or weather. Although the Moon has only one-sixth of Earth's gravity, Armstrong and Aldrin stated that moving around the lunar surface was easier than expected.

Space exploration

APOLLO 13
Manned lunar landing

The seventh of America's Apollo Moon-missions, Apollo 13 almost ended in disaster when an oxygen tank exploded two days after lift-off. As they tried to get the craft back to Earth, the crew had to cope with a lack of water and oxygen. It was a near-fatal situation, but the astronauts finally splashed down safely in the South Pacific.

DATA FILE

SIZE: Command/service module – 11 m (36 ft) long

DATE OF LAUNCH: 11 April 1970

MISSION DURATION: 5 days, 22 hours, 54 minutes, 41 seconds

SPACE AGENCY: NASA

The famous line **"Houston, we've had a problem"** was radioed to Earth by Apollo 13.

The astronauts exited through the crew access hatch

Apollo 13 crew

Following the splashdown in the South Pacific Ocean on 17 April 1970, the command module of the Apollo 13 craft was hoisted aboard the ship USS *Iwo Jima*.

LUNOKHOD 1
Lunar rover

Named after the Russian for "lunar walker", Lunokhod 1 was an unmanned robot rover that, in November 1970, became the first remote-controlled vehicle to visit another world. It landed on the Moon in a basin called Mare Imbrium (the Sea of Showers), took photographs of the lunar surface, examined the Moon's soil, and studied the lunar environment for cosmic rays.

DATA FILE

SIZE: 2.3 m (7.5 ft) tall

DATE OF LAUNCH: 10 November 1970

MISSION DURATION: 10 months, 4 days (last contact)

SPACE AGENCY: Russian space program

The flip-up lid was covered with solar cells

Lunokhod 1 **took more than 20,000 images** of the Moon during its 10-month mission.

VENERA 7
Space probe

The rocket engine propelled the craft to Venus

Venera 7 provided confirmation that **man could not survive on Venus's surface**.

Russian space probe Venera 7 was the first to land on another planet – Venus. It crash-landed after plummeting through the planet's thick atmosphere, but was able to relay data for 23 minutes. It revealed that the atmospheric pressure on Venus was 90 times greater than on Earth's surface, and that the temperature was a searing 475°C (887°F) – hotter than any oven on Earth.

Only the egg-shaped lander went to the surface

CCCP

CC

DATA FILE

SIZE: Lander – 1 m (3 ft) long

DATE OF LAUNCH: 17 August 1970

MISSION DURATION: 3 months, 28 days (last contact)

SPACE AGENCY: Russian space program

SALYUT 1
Space station

Salyut 1 was the first space station to orbit Earth. It was an experiment to discover if astronauts could live and work in space for long periods of time. The first crew on Salyut 1 stayed on board for 23 days – a new record for time spent in space.

DATA FILE

SIZE: 20 m (65.5 ft) long

DATE OF LAUNCH: 19 April 1971

MISSION DURATION: 5 months, 22 days (de-orbited)

SPACE AGENCY: Russian space program

Salyut 1 travelled 118.6 million km (73.7 million miles) in its 175 days in orbit.

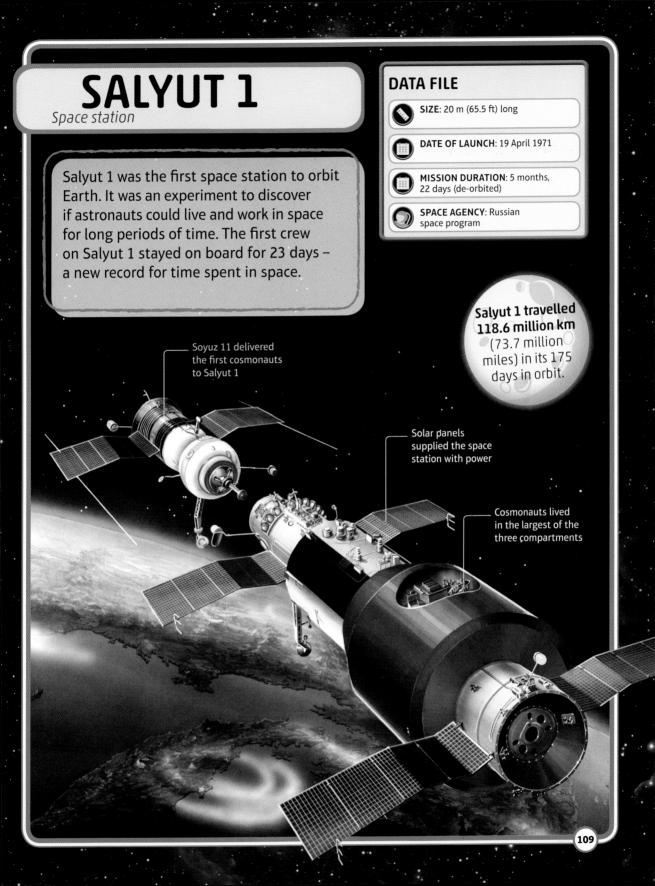

Soyuz 11 delivered the first cosmonauts to Salyut 1

Solar panels supplied the space station with power

Cosmonauts lived in the largest of the three compartments

PIONEER 10
Space probe

Launched on 3 March 1972, Pioneer 10's principal mission was to study Jupiter. On its journey to the gas giant, it became the first spacecraft to cross the Asteroid Belt, dispelling doubts about the risk of collisions in this region. After it had flown past Jupiter, the probe continued to provide information on the most distant parts of our Solar System.

By the time of last contact, Pioneer 10 was **12 billion km** (7.5 billion miles) **away from Earth.**

Antenna relayed data to Earth

Jupiter

DATA FILE

SIZE: 2.9 m (9.5 ft) long

DATE OF LAUNCH: 3 March 1972

MISSION DURATION: 30 years, 10 months, 22 days (at last contact)

SPACE AGENCY: NASA

APOLLO 17
Manned lunar landing

The last of the manned Apollo missions, Apollo 17 was the first to send a scientist into space – the geologist Harrison Schmitt. The astronauts targeted an area on the Moon called Taurus-Littrow, which is rich in volcanic rocks. Lasting for about 75 hours, this was the longest mission on the lunar surface. It also broke the record for the longest moonwalk.

DATA FILE

SIZE: Command module – 11 m (36 ft) long; Rover – 3.1 m (10 ft) long

DATE OF LAUNCH: 7 December 1972

MISSION DURATION: 12 days, 13 hours, 51 minutes

SPACE AGENCY: NASA

Commander Gene Cernan takes the lunar rover out for a ride

Gene Cernan was the last human to set foot on the Moon.

Apollo 17 command/service module

Space exploration

SKYLAB
Space station

Skylab, the United States' first space station, was launched unmanned in 1973. Three separate crews visited it to conduct experiments in physics and astronomy, with missions lasting for months. Skylab was also a laboratory for testing how space crews could cope with weightlessness for long periods of time.

DATA FILE

SIZE: 25.1 m (82.4 ft) long

DATE OF LAUNCH: 14 May 1973

MISSION DURATION: 6 years, 1 month, 27 days (de-orbited)

SPACE AGENCY: NASA

Solar panel

Telescope mount

Command module

Sunshield protects the workshop underneath

In 1979, Skylab was de-orbited and **broke up over Western Australia**.

MARINER 10
Space probe

Mariner 10 flew to within **327 km (203 miles)** of Mercury's surface.

Solar panel

Television cameras

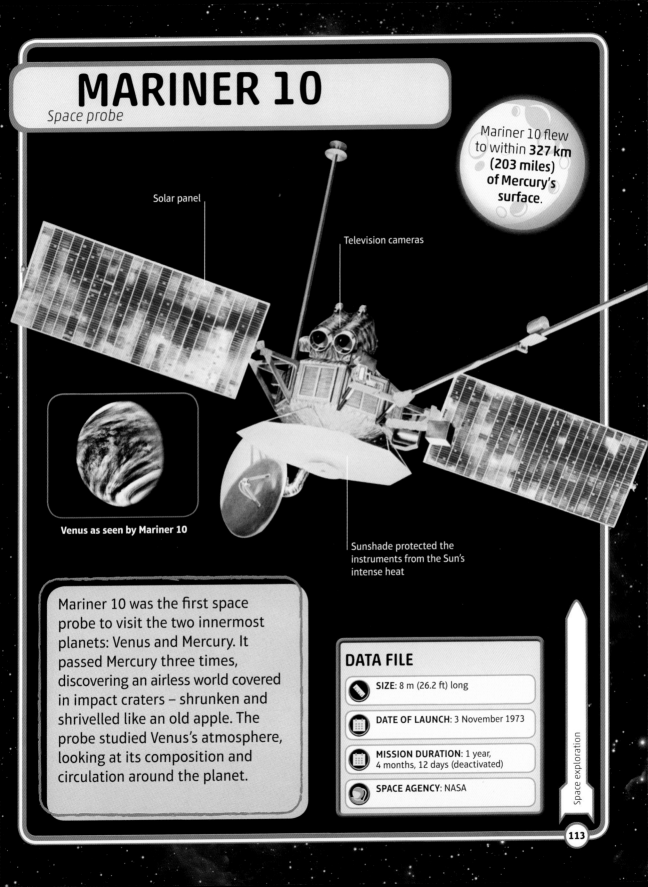

Venus as seen by Mariner 10

Sunshade protected the instruments from the Sun's intense heat

Mariner 10 was the first space probe to visit the two innermost planets: Venus and Mercury. It passed Mercury three times, discovering an airless world covered in impact craters – shrunken and shrivelled like an old apple. The probe studied Venus's atmosphere, looking at its composition and circulation around the planet.

DATA FILE

SIZE: 8 m (26.2 ft) long

DATE OF LAUNCH: 3 November 1973

MISSION DURATION: 1 year, 4 months, 12 days (deactivated)

SPACE AGENCY: NASA

Space exploration

APOLLO-SOYUZ MISSION

Manned spacecraft

The two crews exchanged gifts – including **tree seeds to plant in their respective nations**.

The specially designed docking module allowed the crews to move between the spacecraft

Soyuz 19

UNITED STATES

Apollo command/ service module

The Apollo-Soyuz mission was partly political and partly scientific. The mission saw an Apollo command/service module dock with the Russian craft Soyuz 19, and involved three American and two Russian astronauts. They were docked together for almost two days. The teams collaborated on a number of science experiments.

DATA FILE

SIZE: Combined length of 18.5 m (60 ft)

DATE OF DOCKING: 17 July 1975

MISSION DURATION: 9 days, 1 hour, 28 minutes

SPACE AGENCIES: NASA and Russian space program

Space exploration

VOYAGER 1
Space probe

Space probe Voyager 1 swung past Jupiter on 5 March 1979 and Saturn on 12 November 1980, and sent back sensational pictures of our Solar System's two gas giants. It captured astonishing images of Jupiter's clouds, its Red Spot, and its volcanic moon Io. When the probe reached Saturn, it swept over the planet's glorious rings.

DATA FILE

SIZE: 3.7 m (12 ft) – width of antenna

DATE OF LAUNCH: 5 September 1977

MISSION DURATION: Ongoing

SPACE AGENCY: NASA

Antenna

Voyager 1 has now left our Solar System and **is travelling in interstellar space**.

This system lets the cameras check their colour balance

Great Red Spot as seen by Voyager 1

SPACE SHUTTLE
Reusable manned spacecraft

Until the space shuttle programme, manned missions used vehicles that could only be used once. The concept behind the space shuttle was to make flying into space as easy as taking a plane flight. It was launched by rockets, orbited Earth, then flew down to land on a runway. NASA built five space shuttles, but Challenger and Columbia suffered tragic failures that killed their crews. Discovery, Endeavour, and Atlantis continued to ferry astronauts and supplies to the International Space Station (ISS) until the programme was retired in 2011.

Up to eight astronauts could travel in the crew cabin

DATA FILE

SIZE: 37 m (121 ft) long

NUMBER OF MISSIONS: 135 (between 1981 and 2011)

SPACE AGENCY: NASA

In total, the five NASA space shuttles **flew 873 million km** (542 million miles).

Large parts of the ISS were carried into orbit in the shuttle's cargo bay

WASTE COLLECTION SYSTEM

Weightlessness in space makes going to the toilet very difficult. Space toilets work like a vacuum cleaner, using suction to pull waste away from your body and into a waste tank. Astronauts of both sexes use a hose to collect urine; otherwise they sit down on the toilet.

Waste tank can recycle urine into drinking water

POTABLE AND WASTE TANK

CH2 PRESSURE

The urine collected on Apollo and Gemini craft **was dumped into space**.

Thigh bars prevent the astronaut from floating off the seat

Handle for the vacuum pump that flushes the toilet

PRIOR TO DEFECATOR
1. PULL UP
2. WAIT 5 SEC.
3. PUSH ← PUSH FWD.

COMP
COP

WOE
EMI

ON

OFF

DATA FILE

FIRST USED: 1967 (Soyuz)

SPACE AGENCIES: Russian space program and NASA

SPACESUIT

Astronauts who venture outside their spacecraft need a spacesuit that protects them from the vacuum of space, and also from extreme temperatures and radiation. The suit supplies oxygen to breathe, and – despite its bulkiness – must be flexible enough for astronauts to use their legs, arms, and hands.

Life-support backpack contains oxygen, drinking water, and batteries

Cap with earphones

Checklist of jobs

Gloves with heated rubber fingertips

Battery-powered drill for construction work on a space station

Ventilation garment keeps astronaut cool

The suit has 11 layers; the outer one is a mix of fire-resistant and bullet-proof material

DATA FILE

WEIGHT (ON GROUND): 49 kg (108 lb)

FIRST USED: 18 March 1965 (Alexei Leonov, Voskhod 2)

SPACE AGENCIES: Russian space program and NASA

Space exploration

VENERA 13
Space probe

The spacecraft Venera 13 sent back the first colour pictures from the surface of Venus. Tinted yellow by sunlight shining through clouds, they reveal a barren, volcanic landscape (with parts of the spacecraft in the foreground). The temperature – hotter than an oven – and crushing pressure on Venus destroyed the probe two hours after it had landed.

DATA FILE

WEIGHT: 760 kg (1,675 lb)

DATE OF LAUNCH: 30 October 1981

MISSION DURATION: 4 months, 2 days (last contact)

SPACE AGENCY: Russian space program

The probe survived long enough on Venus's surface to **take 14 photographs**.

Venus's surface as seen by Venera 13

MIR
Space station

In 1990, quail chicks hatched on Mir became the **first animals born in space**.

Cargo supply craft

Core module

Solar panels supplied power, and Mir suffered power cuts in Earth's shadow

The world's first large space station, Mir grew as cosmonauts added extra modules. An international crew – including astronauts from the UK, France, and Japan – carried out many experiments in the weightless environment of Mir. This space station was home to 125 astronauts during its lifetime, until it was de-orbited on 23 March 2001.

DATA FILE

SIZE: 33 m (108 ft)

LAUNCH DATE: 20 February 1986

MISSION DURATION: 15 years, 1 month, 3 days (de-orbited)

SPACE AGENCY: Russian space program

Space exploration

GIOTTO

Space probe

Giotto was the European Space Agency's (ESA) **first deep-space mission**.

The camera took 2,112 images of the comet's nucleus

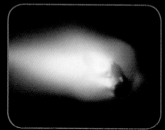

Halley's nucleus as seen by Giotto

When Halley's Comet appeared in 1910, astronomers could only view it with telescopes. In 1986, however, the European Space Agency (ESA) flew a spacecraft called Giotto right through the comet's gaseous head, and took the first pictures of the icy nucleus at its centre. Giotto survived the flight, despite comet-dust hitting it at 100 times the speed of a bullet.

DATA FILE

SIZE: 2.85 m (9.4 ft) long

DATE OF LAUNCH: 2 July 1985

MISSION DURATION: 7 years, 21 days (deactivated)

SPACE AGENCY: ESA

HUBBLE SPACE TELESCOPE

Space telescope

Hubble orbits around Earth at a **speed of 27,400 km/h (17,000 mph)**.

This stunning image taken by the Hubble Space Telescope shows the Antennae galaxies colliding.

Hubble Space Telescope

Orbiting above Earth's atmosphere, Hubble has sharper views of the Universe than any ground-based telescope. During its first 25 years in space, Hubble has made more than 1.2 million amazing observations, ranging from the planets Mars and Jupiter, through stunning nebulae (in which stars are being born), to the most distant galaxies ever seen.

DATA FILE

SIZE: 13.2 m (43 ft) long

DATE OF LAUNCH: 24 April 1990

MISSION DURATION: Ongoing

SPACE AGENCIES: NASA and ESA

Space exploration

GALILEO
Space probe

Controllers ended the mission by **crashing** Galileo **into Jupiter**.

Jupiter's moon Io as seen by Galileo

Named after the great early astronomer, the unmanned Galileo spacecraft was the first to orbit Jupiter – the planet he had studied in the early 17th century. During eight years in orbit, the probe took pictures of Jupiter's clouds, rings, and biggest moons – finding evidence for hidden oceans. A smaller probe launched by Galileo also sampled Jupiter's atmosphere.

DATA FILE

SIZE: 5.3 m (17 ft) long

DATE OF LAUNCH: 18 October 1989

MISSION DURATION: 13 years, 11 months, 3 days (de-orbited)

SPACE AGENCY: NASA

LONG MARCH 3A

Rocket

The Chinese invented firework rockets in the 12th century, but they did not launch their first space rocket until 1970. In the 1990s, they improved the Long March rocket, which has continued to grow ever more powerful. The Long March 3A was designed to launch massive communications satellites into space. In 2007, it launched China's first mission to orbit the Moon.

By mid-2016, Long March 3A rockets had **launched successfully 25 times**.

DATA FILE

SIZE: 52 m (170 ft) tall

FIRST LAUNCH: 8 February 1994

NUMBER OF MISSIONS: 25

SPACE AGENCY: CNSA (China)

ARIANE 5
Rocket

SIZE: 52 m (170 ft) tall

FIRST LAUNCH: 4 June 1996

NUMBER OF MISSIONS: 86 (ongoing)

SPACE AGENCY: ESA

Launched from the French Guiana Space Centre in South America, the European Space Agency's Ariane 5 is a world-leading rocket for launching massive satellites. It has launched European resupply craft to the International Space Station, and will soon send a spacecraft to Mercury. Ariane 5 will launch the James Webb Space Telescope – the successor to Hubble.

The payload fairing protects the satellite during launch

On 1 July 2009, Ariane 5 launched **the largest telecommunications satellite** ever built.

SOJOURNER
Rover

Sojourner's landing on Mars, inside the Mars Pathfinder landing craft, was cushioned by giant airbags as well as a parachute and rockets. The first rover on Mars, Sojourner was designed to last for just a month; in fact, it survived for almost three months, and found that the rocks near the landing site were largely volcanic in origin.

DATA FILE

SIZE: 28 cm (11 in) tall

DATE OF LAUNCH: 4 December 1996

MISSION DURATION: 9 months, 23 days (last contact)

SPACE AGENCY: NASA

Sojourner

The rover was **named after Sojourner Truth**, a pioneer of women's rights.

Space exploration

INTERNATIONAL SPACE STATION

Space station

Orbiting 400 km (249 miles) above Earth's surface, the International Space Station (ISS) is the biggest man-made object in space. The individual modules were sent on American space shuttles or Russian Soyuz craft, and were assembled by space-walking astronauts. The crew conducts many experiments on board, including studying the effects of weightlesness on the human body over long periods, to prepare for manned trips to Mars.

Japanese science laboratory Kibo

Canadian robotic arm helps move equipment

European laboratory Columbus

American module Harmony has closets that work as bedrooms

DATA FILE

SIZE: 108.5 m (356 ft) long

FIRST MODULE LAUNCHED: 20 November 1998

MISSION DURATION: Ongoing

SPACE AGENCIES: NASA, ROSCOSMOS (Russia), JAXA (Japan), ESA, CSA (Canada)

The ISS has been **occupied continuously** since November 2000.

ASSEMBLING THE ISS

Zarya, 1998
The first module, Zarya was launched on a Russian rocket in 1998. It was soon joined by the American module, Unity.

Zvezda, 2000
Launched in 2000, the Russian module Zvezda was the third. It provided life support for the first crew on the ISS.

Laboratories, 2005
In 2005, the space shuttle Discovery had a grandstand view of the half-finished ISS when it delivered scientific equipment to the station.

Completed, 2011
The final space shuttle flight in 2011 saw Atlantis travel to the ISS. The space station was now complete.

Mating adapters allow spacecraft and modules to dock

Russian module Zarya

Soyuz spacecraft, used to take astronauts to and from the ISS

The Italian-built Leonardo module is a storage area

Russian module Zvezda

CHANDRA X-RAY OBSERVATORY
Space telescope

The outer clouds of this dying star give it the name Cat's Eye Nebula

Chandra was **named after Subrahmanyan Chandrasekhar**, a 20th-century astrophysicist.

Chandra has shown that a cloud of multi-million degree gas surrounds the central star

Artist's impression of Chandra

Launched on 23 July 1999, the Chandra observatory studies X-rays – very high-energy radiation – from the distant Universe. It has discovered giant clouds of hot gas held together by mysterious dark matter, and has also examined streams of gas as they fall into black holes.

DATA FILE

SIZE: 19.5 m (64 ft) long

DATE OF LAUNCH: 23 July 1999

MISSION DURATION: Ongoing

SPACE AGENCY: NASA

Space exploration

NEAR SHOEMAKER
Space probe

The Near Earth Asteroid Rendezvous (NEAR) spacecraft was renamed NEAR Shoemaker after Gene Shoemaker, a scientist who investigated craters on Earth that had been blasted out by asteroids. Designed to study asteroid Eros, it became the first space mission to orbit an asteroid and to land on its surface. It found Eros to be a peanut-shaped world, scarred with over 100,000 craters.

DATA FILE

SIZE: 2.7 m (9 ft) long

DATE OF LAUNCH: 17 February 1996

MISSION DURATION: 5 years, 12 days (last contact)

SPACE AGENCY: NASA

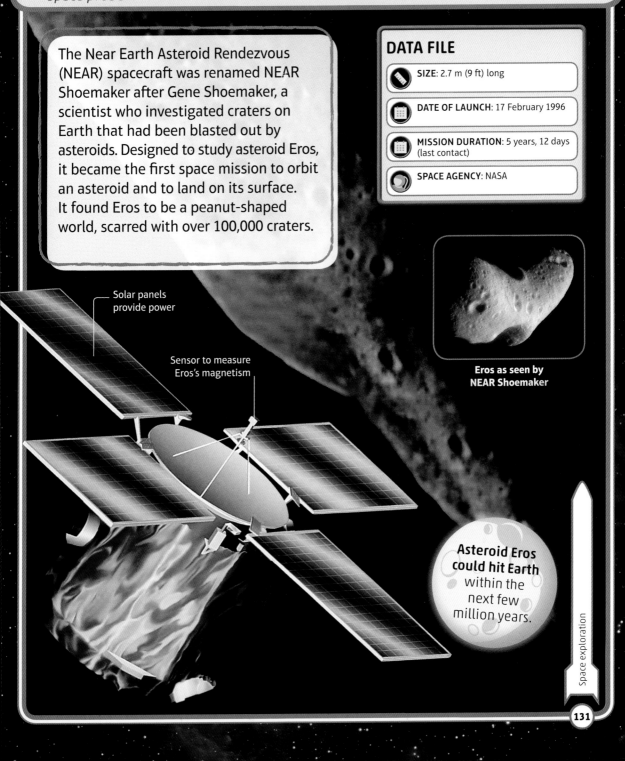

Eros as seen by NEAR Shoemaker

Solar panels provide power

Sensor to measure Eros's magnetism

Asteroid Eros could hit Earth within the next few million years.

SPACESHIPONE

Reusable manned spacecraft

SpaceShipOne, piloted by Mike Melville, reached space on 21 June 2004, becoming the first privately built spacecraft to do so. It was launched from the carrier plane White Knight at an altitude of 14 km (9 miles) before firing its own rockets to propel it beyond Earth's atmosphere. Virgin Galactic is building a larger version, SpaceShipTwo, which will carry six paying "space tourists".

Vertical booms and horizontal stabilizers on the wings helped the spacecraft remain stable when re-entering the atmosphere

SpaceShipOne was only designed **to** carry a pilot and two customers.

N328KF

DATA FILE

SIZE: 8.5 m (28 ft) long

FIRST MANNED FLIGHT TO SPACE: 21 June 2004

NUMBER OF MISSIONS: 3 (last flight 4 October 2004)

MANUFACTURER: Scaled Composites (private company)

The fuselage and wings are made of lightweight carbon-based materials

OPPORTUNITY
Rover

Opportunity landed on Mars on 25 January 2004 and has travelled over 42 km (26 miles) over the Red Planet's deserts since then, studying Martian geology. It has proved that Mars once had flowing liquid water, even though it is now cold and dry. Opportunity has survived being stuck in the sand and a terrible Martian dust storm.

DATA FILE

SIZE: 1.5 m (5 ft) tall

DATE OF LAUNCH: 7 July 2003

MISSION DURATION: Ongoing

SPACE AGENCY: NASA

Designed to last 90 days, **Opportunity has survived for over 12 years**.

Image of Mars's surface taken by Opportunity

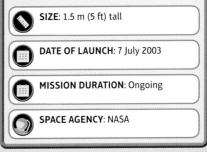

Artist's impression of Opportunity on the surface of Mars

Space exploration

CASSINI-HUYGENS
Orbiter and lander

The main radio dish on the Cassini probe is 4 m (13 ft) across

Cassini-Huygens was the first mission to orbit Saturn. Orbiter Cassini carried the small lander Huygens, which parachuted down through the clouds of the planet's biggest moon Titan and landed on its frozen plains. Cassini has taken close-up images of Saturn, its rings, and many of its icy moons.

The Cassini orbiter **took six-and-a-half years to reach Saturn.**

DATA FILE

SIZE: 6.8 m (22 ft) long

DATE LAUNCH: 15 October 1997

MISSION DURATION: Ongoing

SPACE AGENCIES: NASA and ESA

Cassini's view of Saturn

HAYABUSA
Unmanned spacecraft

In September 2005, Hayabusa reached the asteroid Itokawa, about 300 million km (186.4 million miles) away from Earth. It landed on the asteroid's surface and collected dust grains. It was the first mission to bring asteroid samples back to Earth. After analysing the dust grains, scientists found that Itokawa had been part of a larger asteroid that broke up millions of years ago.

DATA FILE

SIZE: 1.5 m (5 ft) wide (without solar panels)

DATE OF LAUNCH: 9 May 2003

MISSION DURATION: 7 years, 1 month, 4 days

SPACE AGENCY: JAXA (Japan)

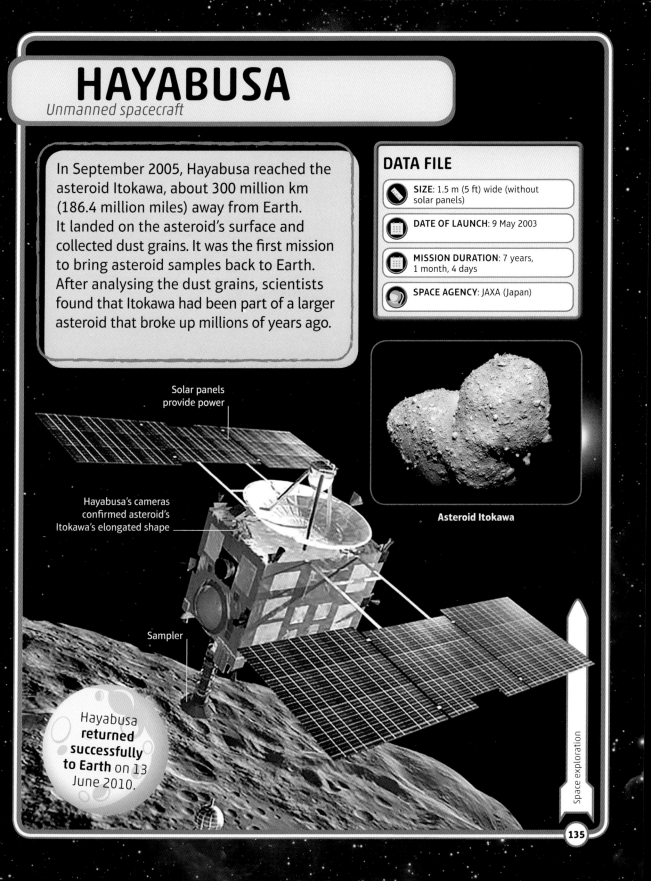

Solar panels provide power

Hayabusa's cameras confirmed asteroid's Itokawa's elongated shape

Asteroid Itokawa

Sampler

Hayabusa **returned successfully to Earth** on 13 June 2010.

KEPLER
Space telescope

Launched in March 2009, the Kepler Space Telescope has discovered a record number of planets orbiting other stars. It has spent years studying a small region in the constellations Cygnus and Lyra, checking for stars that dim periodically as planets move in front of them. By mid-2016, Kepler had discovered 2,325 planets.

DATA FILE

SIZE: 4.7 m (15.4 ft) long

DATE OF LAUNCH: 7 March 2009

MISSION DURATION: Ongoing

SPACE AGENCY: NASA

Sun shade protects mirror from the Sun's glare

Star tracker

Body of telescope

Radiator to cool telescope

The telescope was **named after** the 16th-century astronomer, **Johannes Kepler**.

CURIOSITY
Rover

DATA FILE

SIZE: 2.2 m (7.2 ft) tall

DATE OF LAUNCH: 26 November 2011

MISSION DURATION: Ongoing

SPACE AGENCY: NASA

Curiosity landed on Mars on 6 August 2012, with a mission to explore the Red Planet's surface. The largest and most complex rover ever sent to another planet, it has found organic molecules – the building blocks of life – in Martian rocks. It has also sniffed traces of methane gas, which can be produced by living cells.

Laser vapourizes rocks up to 7 m (23 ft) away, to analyse their composition

The Martian surface as seen by Curiosity

Curiosity played **"Happy Birthday to You"** on the first anniversary of its landing.

ROSETTA AND PHILAE
Orbiter and lander

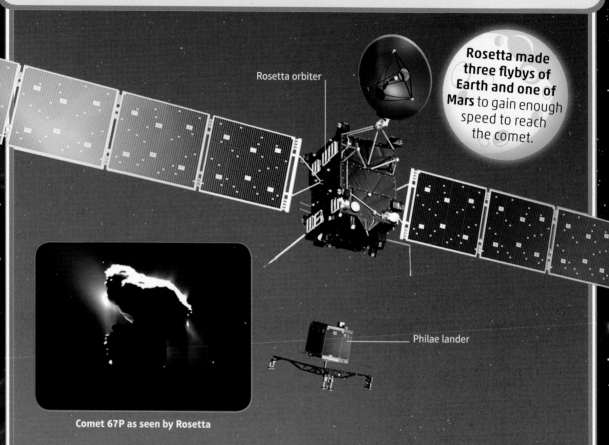

Rosetta orbiter

Rosetta made three flybys of Earth and one of Mars to gain enough speed to reach the comet.

Comet 67P as seen by Rosetta

Philae lander

The European Space Agency's Rosetta mission was the first in history to meet up with a comet – Comet 67P – and to follow it as it orbited the Sun. On 12 November 2014, the lander Philae detached from Rosetta and became the first man-made object to land on a comet's surface when it touched down on 67P.

DATA FILE

SIZE: Rosetta – 2.8 m (9 ft) wide (without solar panels); Philae – 1 m (3.3 ft) wide

DATE OF LAUNCH: 2 March 2004

MISSION DURATION: 12 years, 6 months

SPACE AGENCY: ESA

NEW HORIZONS
Space probe

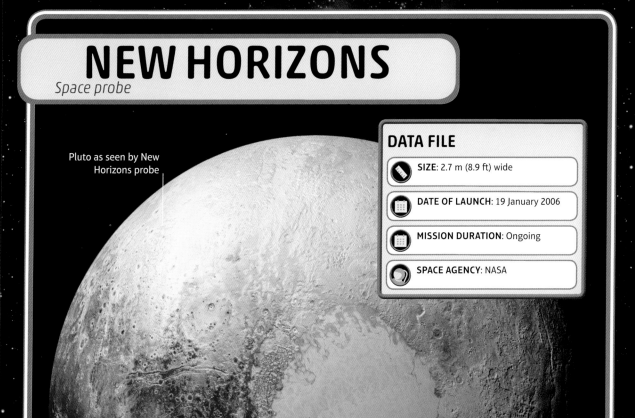

Pluto as seen by New Horizons probe

DATA FILE

SIZE: 2.7 m (8.9 ft) wide

DATE OF LAUNCH: 19 January 2006

MISSION DURATION: Ongoing

SPACE AGENCY: NASA

New Horizons raced through space at a **speed of up to 14.5 km (9 miles) per second**.

When New Horizons was launched in January 2006, its destination – Pluto – was the ninth planet of our Solar System, but by the time it arrived in July 2015, the tiny globe had been demoted to the status of a dwarf planet. The space probe has discovered Pluto to be a strange world, with mountains of ice and smooth plains of frozen nitrogen and carbon monoxide.

Artist's impression of the New Horizons probe

Space exploration

FALCON 9
Reusable rocket

Dragon capsule delivers payloads such as satellites into space

Falcon 9 was the first powerful, privately built rocket. With its Dragon capsule, it pioneered the commercial delivery of cargo to the International Space Station (ISS). The first stage of the Falcon 9 returns to Earth and can be reused in future flights, while the Dragon capsule is now being upgraded to carry astronauts into space.

On 8 April 2016, Falcon 9 completed the **first vertical landing by a space rocket**.

First-stage engines burn kerosene and liquid oxygen

DATA FILE

SIZE: 70 m (230 ft) tall

FIRST LAUNCH: 4 June 2010

NUMBER OF MISSIONS: 25

ORGANIZATION: SpaceX (private company)

SPACE LAUNCH SYSTEM

Rocket

The escape tower will pull the Orion capsule clear in case of an emergency

Orion capsule inside here

The Space Launch System (SLS), NASA's new rocket, will be the most powerful ever built. Powered by the same engines used in the space shuttle, with similar boosters, it is not reusable. Its Orion capsule has been designed to carry a crew of four astronauts beyond Earth orbit, to land on an asteroid or the Moon and – eventually – to travel to Mars.

The solid-rocket boosters fire for 126 seconds before being jettisoned

The SLS **will be launched from the Kennedy Space Centre** in Florida, USA.

DATA FILE

 SIZE: 65 m (213 ft) tall

PLANNED DATE OF LAUNCH: November 2018

SPACE AGENCY: NASA

Space exploration

5
STAR MAPS AND STARGAZING

Finding your way around the stars at night is like exploring a foreign country: it may seem confusing at first, but with the help of a good map you can navigate the sky. As well as identifying the stars and fainter objects (such as clusters, nebulae, and galaxies), it is also fascinating to discover the history of the constellation names.

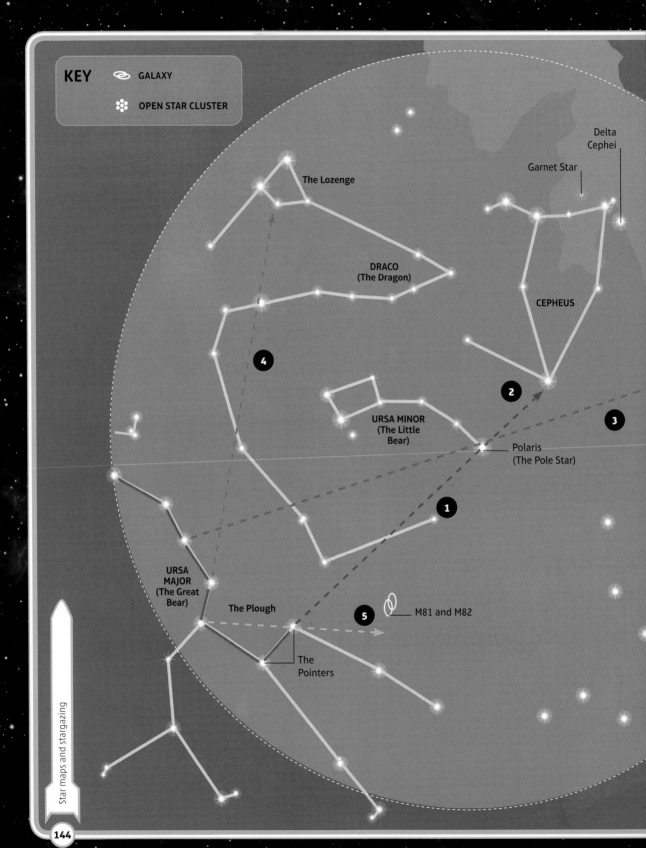

KEY

◎ GALAXY

✳ OPEN STAR CLUSTER

The Lozenge

DRACO
(The Dragon)

Delta Cephei

Garnet Star

CEPHEUS

4

URSA MINOR
(The Little Bear)

2

Polaris
(The Pole Star)

3

1

URSA MAJOR
(The Great Bear)

The Plough

5

M81 and M82

The Pointers

STARGAZING BASICS
Northern star hopping

To find your way around the night sky, it is easiest to pick out some obvious landmarks and then trace imaginary lines from these to other constellations. This technique is called star hopping and is easy to do with the naked eye, though you can see more if you have binoculars. The chart on this page shows how to star-hop around Polaris, the Pole Star, which is visible to people who live in Earth's northern hemisphere.

CASSIOPEIA

NGC 457

The hazy cloud of the Milky Way

Sailors used the Pole Star **to find North** on a clear night.

FINDING THE WAY

1 Find the Plough (in red) and locate the two stars furthest from its "handle", which are known as the Pointers. Draw an imaginary line (purple) through the Pointers and extend it to a star of similar brightness. This is the Pole Star and it is always due north.

2 Extend the line from the Pointers past the Pole Star to reach the constellation Cepheus, which looks a bit like a lopsided house. Binoculars will reveal the bright red Garnet Star at the base of the house – the reddest star visible to the naked eye.

3 Now draw a line (green) from the third star down the Plough's handle through the Pole Star to find the constellation Cassiopeia, which looks like a flattened "W". If you have binoculars, look for a star cluster just below the central peak of the "W".

4 The large but faint constellation Draco (the Dragon) is best seen under very dark skies. A line (pink) from the fourth star down the Plough's handle cuts across the dragon's body and carries on to its head, a pattern of stars called the Lozenge.

5 Track down one of the brightest galaxies in the sky with binoculars: follow a diagonal line (orange) across the Plough's rectangle of stars, and continue in the same direction looking for a pair of tiny fuzzy patches. These are the galaxies M81 and its fainter companion M82.

Star maps and stargazing

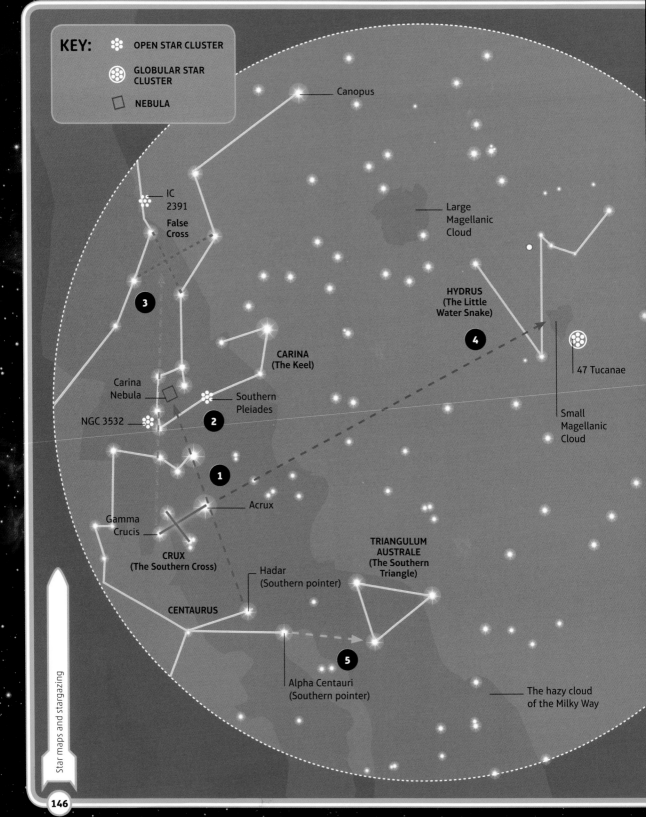

KEY:

- ⚬ OPEN STAR CLUSTER
- ⚬ GLOBULAR STAR CLUSTER
- ▢ NEBULA

Canopus

IC 2391

False Cross

Large Magellanic Cloud

HYDRUS (The Little Water Snake)

4

3

CARINA (The Keel)

47 Tucanae

Carina Nebula

Southern Pleiades

2

Small Magellanic Cloud

NGC 3532

1

Acrux

Gamma Crucis

TRIANGULUM AUSTRALE (The Southern Triangle)

CRUX (The Southern Cross)

Hadar (Southern pointer)

CENTAURUS

Alpha Centauri (Southern pointer)

5

The hazy cloud of the Milky Way

STARGAZING BASICS
Southern star hopping

This chart shows you how to star-hop your way around some of the top sights in the southern night sky, visible to people who live in Earth's southern hemisphere. The southern sky gives stargazers a fantastic view of our Milky Way galaxy, as well as celestial wonders such as colourful nebulae, whole galaxies, and the bright constellations Carina, Centaurus, and the Southern Cross.

FINDING THE WAY

1 Identify Crux (in red), the Southern Cross (not to be confused with the False Cross) and the Southern Pointer stars Alpha Centauri and Hadar. Draw a line (green) from Hadar to the bottom of the Southern Cross and carry on for the same distance again to reach the famous Carina Nebula.

2 Two beautiful star clusters lie close to the Carina Nebula: NGC 3532 and the Southern Pleiades (IC 2602). This second cluster contains five or six naked-eye stars – see how many you can count (you will see more by looking slightly to one side of it). Then use binoculars to view many more.

3 Next follow a line (pink) from the top of the Southern Cross, past the Carina Nebula, and onwards by the same distance again. Here you will find the deceptive pattern of the False Cross, and just beyond it the star cluster IC 2391. This impressive jewel box of stars is best appreciated through binoculars.

4 Now follow (purple) the downward (longer) bar of the Southern Cross and cross an empty area of the sky to reach the Small Magellanic Cloud. This small galaxy orbits our own Milky Way galaxy and contains hundreds of millions of stars. Nearby is an impressive globular cluster of stars known as 47 Tucanae.

5 Finally, return to the Southern Pointers and follow a line (orange) from Alpha Centauri to discover three bright stars that form a triangle shape – named constellation Triangulum Australe.

The False Cross **mimics the shape of the true Southern Cross**, but is slightly fainter.

Star maps and stargazing

THE NORTHERN SKY

From Earth about 6,000 stars are visible to the naked eye in the night sky, though you can only see about half of these from any location at any one time. Ancient astronomers saw the night sky as a giant sphere of stars, or celestial sphere, surrounding Earth. The stars and constellations of the northern half of this sphere are shown on this flat sky map.

DELPHINUS

AQUILA

SAGITTA

VULPECULA

SERPENS CAUDA

KEY

* YELLOW STAR

* RED STAR

* ORANGE STAR

* WHITE STAR

* BLUE STAR

OPHIUCHUS

Polaris, or the Pole Star, **remains directly over** Earth's north pole.

HERCULES

SERPENS CAPUT

NORTHERN SKY

Most of the constellation names in the northern hemisphere come from the ancient Greeks. They are often linked to myths, such as the story of Perseus and Andromeda, but some of the fainter stars lie in constellations with more modern names.

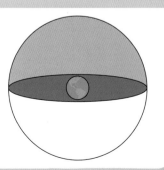

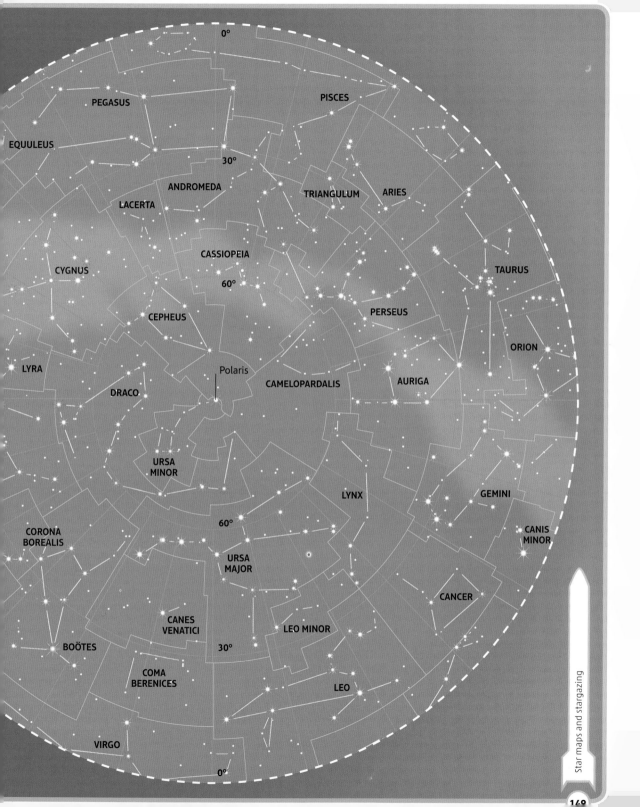

0°

PEGASUS

PISCES

EQUULEUS

30°

ANDROMEDA

TRIANGULUM   ARIES

LACERTA

CASSIOPEIA

CYGNUS

60°

TAURUS

CEPHEUS

PERSEUS

ORION

LYRA

Polaris

DRACO

CAMELOPARDALIS   AURIGA

URSA
MINOR

LYNX

GEMINI

60°

CANIS
MINOR

CORONA
BOREALIS

URSA
MAJOR

CANCER

CANES
VENATICI

LEO MINOR

BOÖTES

30°

COMA
BERENICES

LEO

VIRGO

0°

THE SOUTHERN SKY

As Earth orbits the Sun, different parts of the celestial sphere appear above us, which means we see a changing sequence of constellations over a year. The stars and constellations of the southern half of the celestial sphere are shown in this flat sky map. Earth's south pole lines up with the centre of this map.

KEY

* YELLOW STAR
* RED STAR
* ORANGE STAR
* WHITE STAR
* BLUE STAR

The stars near the edge of this map **can also be seen** from the northern hemisphere.

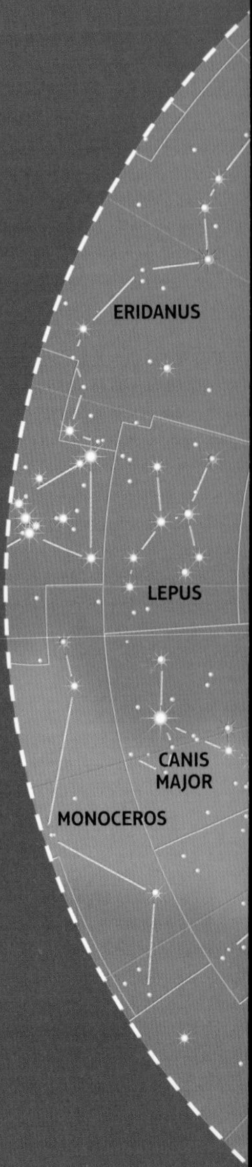

ERIDANUS

LEPUS

CANIS MAJOR

MONOCEROS

SOUTHERN SKY

Southern-hemisphere stars close to the equator (around the edge of the map) were visible to ancient Greek astronomers who grouped them into mythological constellations. Names for constellations around the south celestial pole were proposed by astronomers working from the late 16th century onwards.

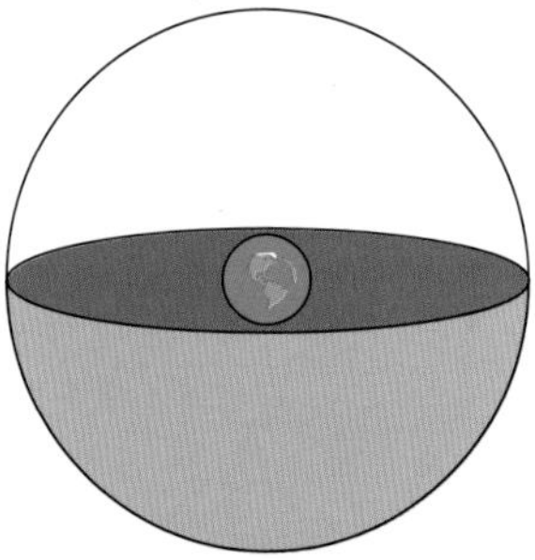

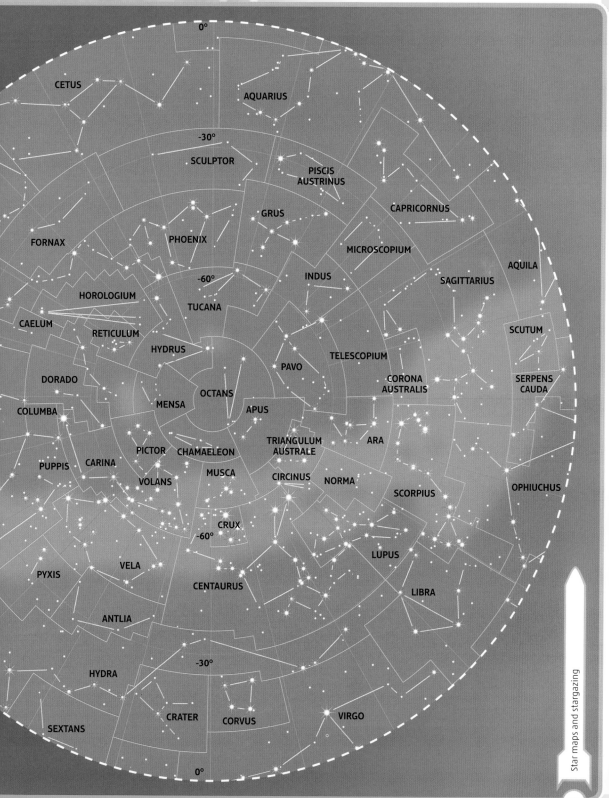

CETUS
AQUARIUS
0°
-30°
SCULPTOR
PISCIS
AUSTRINUS
CAPRICORNUS
GRUS
FORNAX
PHOENIX
MICROSCOPIUM
AQUILA
-60°
INDUS
SAGITTARIUS
HOROLOGIUM
TUCANA
CAELUM
RETICULUM
SCUTUM
HYDRUS
TELESCOPIUM
SERPENS
CAUDA
PAVO
DORADO
CORONA
AUSTRALIS
OCTANS
COLUMBA
MENSA
APUS
PUPPIS
CARINA
PICTOR
CHAMAELEON
TRIANGULUM
AUSTRALE
ARA
OPHIUCHUS
VOLANS
MUSCA
CIRCINUS
NORMA
SCORPIUS
CRUX
-60°
LUPUS
PYXIS
VELA
CENTAURUS
LIBRA
ANTLIA
HYDRA
-30°
VIRGO
SEXTANS
CRATER
CORVUS
0°

URSA MINOR
The Little Bear

Ursa Minor is home to Polaris, the North Star. It lies over the north pole, making the star a fixed point in the sky over Earth's northern hemisphere.

DATA FILE

SIZE RANKING: 56th

BRIGHTEST STARS: Polaris and Kochab

MONTH WHEN BEST VISIBLE: June – Northern hemisphere

Polaris

Kochab

DRACO
The Dragon

Draco, a constellation that represents a dragon, is sandwiched between Ursa Major and Ursa Minor. Its brightest star, Eltanin, lies in the dragon's head.

Eltanin

Aldibain

DATA FILE

SIZE RANKING: 8th

BRIGHTEST STARS: Eltanin and Aldibain

MONTH WHEN BEST VISIBLE: July – Northern hemisphere

CEPHEUS

Cepheus is named after a king of Greek mythology. Its most famous star is Delta Cephei, which swells and shrinks, and has inspired the term "Cepheid" – used for stars that display this behaviour.

DATA FILE

- **SIZE RANKING:** 27th
- **BRIGHTEST STAR:** Alderamin
- **MONTH WHEN BEST VISIBLE:** October – Northern hemisphere

Alderamin

Delta Cephei

CASSIOPEIA

W-shaped Cassiopeia, named after the wife of King Cepheus, is home to the remains of an exploding star, a supernova, which blew up in 1572. It can be seen with a powerful telescope.

DATA FILE

- **SIZE RANKING:** 25th
- **BRIGHTEST STARS:** Shedar and Gamma Cassiopeiae
- **MONTH WHEN BEST VISIBLE:** November – Northern hemisphere

Gamma Cassiopeiae

Shedar

CAMELOPARDALIS
The Giraffe

The Greeks called giraffes "camel leopards" because of their long necks and spotted bodies, and this is how Camelopardalis gets its name. It is a very faint constellation.

DATA FILE

SIZE RANKING: 18th

BRIGHTEST STAR: Beta Camelopardalis

MONTH WHEN BEST VISIBLE:
February – Northern hemisphere

Beta Camelopardalis

AURIGA
The Charioteer

The constellation Auriga is home to Capella, the sixth-brightest star in the sky. To the ancient Greeks, the constellation represented a charioteer carrying a goat and two baby goats on his arm.

Capella

DATA FILE

SIZE RANKING: 21st

BRIGHTEST STAR: Capella

MONTH WHEN BEST VISIBLE:
February – Northern hemisphere

LYNX

Polish astronomer Johannes Hevelius created this extremely faint constellation in 1687. He drew a lynx around the stars because he said that you would need the eyesight of a lynx to see it.

DATA FILE

- **SIZE RANKING:** 28th
- **BRIGHTEST STAR:** Alpha Lyncis
- **MONTH WHEN BEST VISIBLE:** March – Northern hemisphere

Alpha Lyncis

URSA MAJOR
The Great Bear

The seven main stars of Ursa Major are known as "The Plough", or the "Big Dipper" (shown in purple). Mizar, in the "bear's tail", has a companion star called Alcor that is visible to the naked eye.

Dubhe

Alcor Alioth

Mizar

The Plough

DATA FILE

- **SIZE RANKING:** 3rd
- **BRIGHTEST STARS:** Alioth and Dubhe
- **MONTH WHEN BEST VISIBLE:** April – Northern hemisphere

CANES VENATICI
The Hunting Dogs

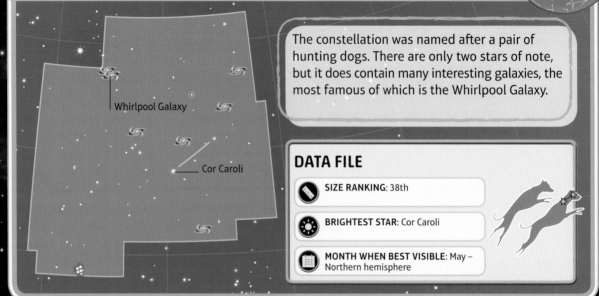

Whirlpool Galaxy

Cor Caroli

The constellation was named after a pair of hunting dogs. There are only two stars of note, but it does contain many interesting galaxies, the most famous of which is the Whirlpool Galaxy.

DATA FILE

SIZE RANKING: 38th

BRIGHTEST STAR: Cor Caroli

MONTH WHEN BEST VISIBLE: May – Northern hemisphere

BOÖTES
The Herdsman

Boötes represents a man herding the Great Bear around the pole. It includes the brilliant red giant Arcturus, the brightest star in the northern sky.

DATA FILE

SIZE RANKING: 13th

BRIGHTEST STAR: Arcturus

MONTH WHEN BEST VISIBLE: June – Northern hemisphere

Arcturus

HERCULES

This constellation is named after Hercules, the strong man of Greek mythology. It is faint, but it hosts M13, a globular cluster of about half a million stars that is visible to the naked eye.

DATA FILE

- **SIZE RANKING:** 5th
- **BRIGHTEST STAR:** Kornephoros
- **MONTH WHEN BEST VISIBLE:** July – Northern hemisphere

M13

Kornephoros

LYRA
The Lyre

Lyra boasts Vega, the fifth-brightest star in the sky. The Ring Nebula – the faint remains of a star that puffed off its atmosphere in its death throes – is visible through a telescope.

DATA FILE

- **SIZE RANKING:** 52nd
- **BRIGHTEST STAR:** Vega
- **MONTH WHEN BEST VISIBLE:** August – Northern hemisphere

Vega

Ring Nebula

CYGNUS
The Swan

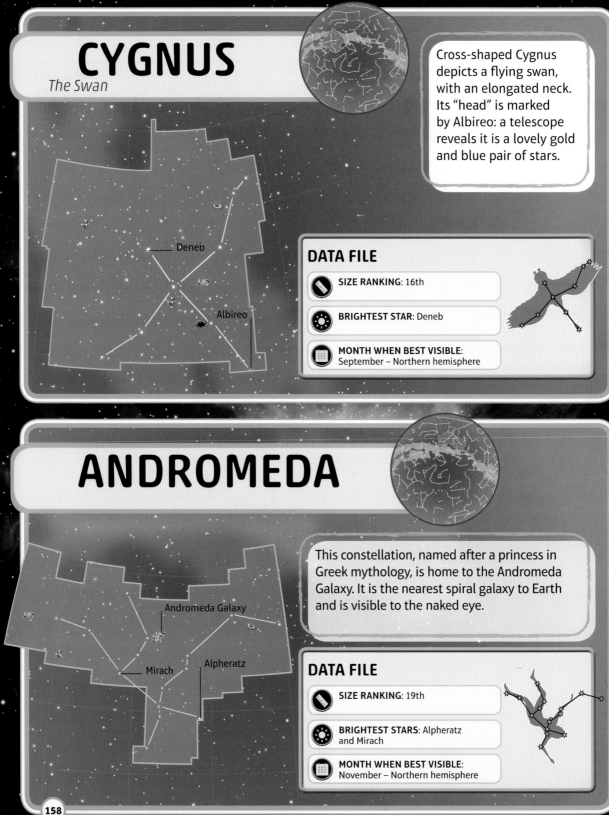

Cross-shaped Cygnus depicts a flying swan, with an elongated neck. Its "head" is marked by Albireo: a telescope reveals it is a lovely gold and blue pair of stars.

Deneb

Albireo

DATA FILE

SIZE RANKING: 16th

BRIGHTEST STAR: Deneb

MONTH WHEN BEST VISIBLE:
September – Northern hemisphere

ANDROMEDA

This constellation, named after a princess in Greek mythology, is home to the Andromeda Galaxy. It is the nearest spiral galaxy to Earth and is visible to the naked eye.

Andromeda Galaxy

Mirach

Alpheratz

DATA FILE

SIZE RANKING: 19th

BRIGHTEST STARS: Alpheratz and Mirach

MONTH WHEN BEST VISIBLE:
November – Northern hemisphere

LACERTA
The Lizard

Polish astronomer Johannes Hevelius created this dim, sprawling constellation in 1697. It represents a lizard scuttling between Andromeda and Cygnus.

Alpha Lacertae

DATA FILE

SIZE RANKING: 68th

BRIGHTEST STAR: Alpha Lacertae

MONTH WHEN BEST VISIBLE:
October – Northern hemisphere

TRIANGULUM
The Triangle

The jewel in this constellation is the stunning Triangulum Galaxy. Just half the size of the Milky Way, this spiral galaxy is the third-largest member of our local group of galaxies.

Deltoton

Triangulum Galaxy

DATA FILE

SIZE RANKING: 78th

BRIGHTEST STAR: Deltoton

MONTH WHEN BEST VISIBLE:
December – Northern hemisphere

PERSEUS

Perseus was a hero of Greek mythology who cut off the head of the evil Medusa. It is marked in the sky by Algol, which "winks" as one star regularly hides its companion.

Mirfak

Algol

DATA FILE

SIZE RANKING: 24th

BRIGHTEST STAR: Mirfak

MONTH WHEN BEST VISIBLE:
December – Northern hemisphere

ARIES
The Ram

Aries represents a ram with a golden fleece in Greek mythology. The constellation's most obvious feature is a crooked line of three stars south of Triangulum, the brightest star of which is Hamal.

Hamal

DATA FILE

SIZE RANKING: 39th

BRIGHTEST STAR: Hamal

MONTH WHEN BEST VISIBLE:
December – Northern hemisphere

TAURUS
The Bull

Taurus is one of the most interesting constellations. The bull's "head" is drawn around the Hyades star cluster, and the red giant Aldebaran is its "eye". Taurus also boasts the Crab Nebula, the remains of an exploded star.

Crab Nebula

Aldebaran

Hyades

DATA FILE

SIZE RANKING: 17th

BRIGHTEST STAR: Aldebaran

MONTH WHEN BEST VISIBLE:
January – Northern hemisphere

GEMINI
The Twins

The two brightest stars in this constellation are named after the mythical twins Castor and Pollux. Amazing Castor is actually a family of six stars. Pollux is circled by a planet bigger than the gas giant Jupiter.

Castor

Pollux

DATA FILE

SIZE RANKING: 30th

BRIGHTEST STARS: Pollux and Castor

MONTH WHEN BEST VISIBLE:
February – Northern hemisphere

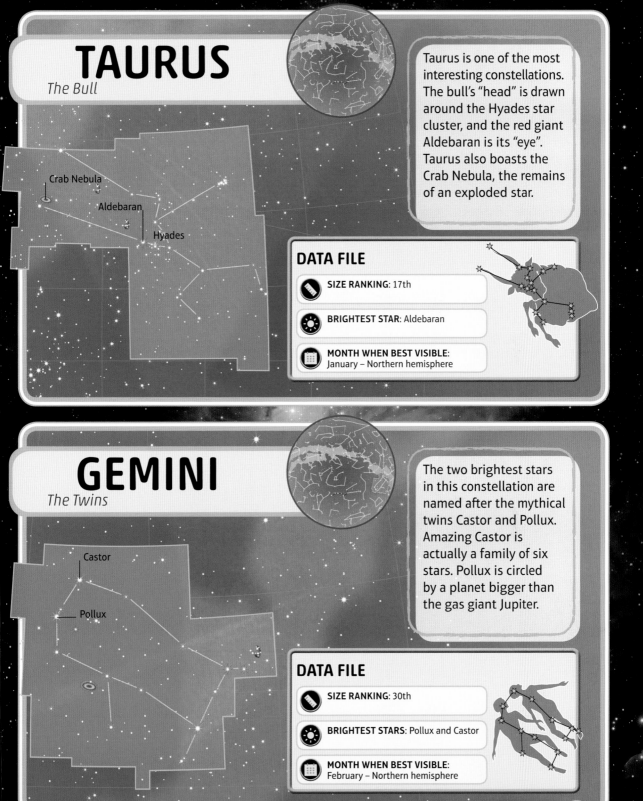

CANCER
The Crab

The faintest constellation of the Zodiac (the band of sky through which the Sun moves), Cancer contains the spectacular star cluster Praesepe, which looks like a swarm of bees.

DATA FILE

SIZE RANKING: 31st

BRIGHTEST STAR: Altarf

MONTH WHEN BEST VISIBLE:
March – Northern hemisphere

Praesepe

Altarf

LEO MINOR
The Little Lion

This faint constellation represents a small lion, on the back of Leo (the lion). Leo Minor was introduced by Johannes Hevelius in 1687.

Praecipua

DATA FILE

SIZE RANKING: 64th

BRIGHTEST STAR: Praecipua

MONTH WHEN BEST VISIBLE: April – Northern hemisphere

COMA BERENICES
Berenices' Hair

Coma Berenices is a faint, but interesting, constellation. Dozens of faint stars form a wedge-shaped group called the Coma Star Cluster, easily seen through binoculars.

DATA FILE

- **SIZE RANKING**: 42nd
- **BRIGHTEST STAR**: Beta Comae Berenices
- **MONTH WHEN BEST VISIBLE**: May – Northern hemisphere

Beta Comae Berenices

Coma star cluster

LEO
The Lion

Leo really looks like what it is supposed to represent – a crouching lion. Its brightest star, Regulus, marks Leo's heart, and spins in just 16 hours – compared to the Sun's 30 days.

Regulus

DATA FILE

- **SIZE RANKING**: 12th
- **BRIGHTEST STAR**: Regulus
- **MONTH WHEN BEST VISIBLE**: April – Northern hemisphere

VIRGO
The Virgin

The glory of this constellation lies in its "bowl". Using a telescope it is possible to see some of the 2,000 galaxies making up the giant Virgo Cluster, which lies about 55 million light years away.

Virgo Cluster

Spica

DATA FILE

SIZE RANKING: 2nd

BRIGHTEST STAR: Spica

MONTH WHEN BEST VISIBLE: May – Northern hemisphere

LIBRA
The Scales

The scales of justice are represented by Libra, and are held by the goddess of justice, Virgo. In ancient times, the stars in this area of sky were thought to make up the claws of Scorpius.

Zubeneschamali

DATA FILE

SIZE RANKING: 29th

BRIGHTEST STAR: Zubeneschamali

MONTH WHEN BEST VISIBLE: June – Southern hemisphere

CORONA BOREALIS
The Northern Crown

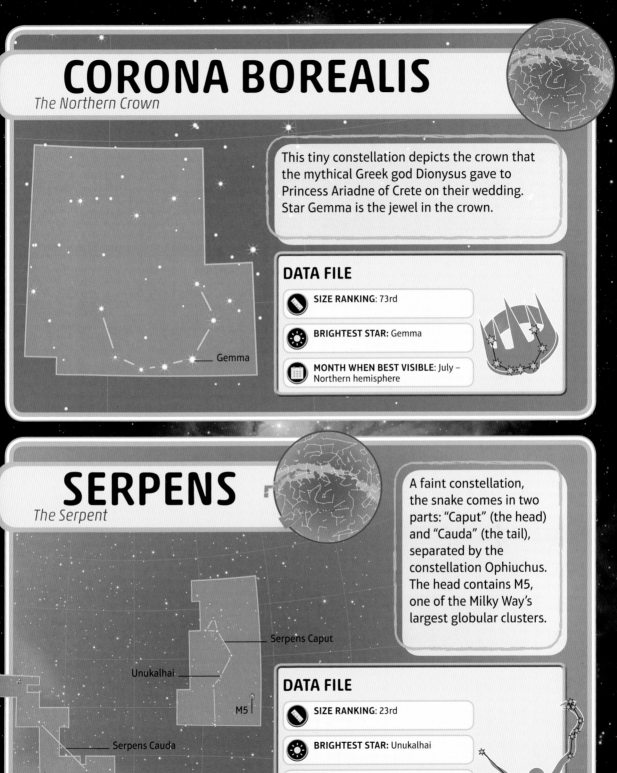

This tiny constellation depicts the crown that the mythical Greek god Dionysus gave to Princess Ariadne of Crete on their wedding. Star Gemma is the jewel in the crown.

Gemma

DATA FILE

SIZE RANKING: 73rd

BRIGHTEST STAR: Gemma

MONTH WHEN BEST VISIBLE: July – Northern hemisphere

SERPENS
The Serpent

A faint constellation, the snake comes in two parts: "Caput" (the head) and "Cauda" (the tail), separated by the constellation Ophiuchus. The head contains M5, one of the Milky Way's largest globular clusters.

Serpens Caput

Unukalhai

M5

Serpens Cauda

DATA FILE

SIZE RANKING: 23rd

BRIGHTEST STAR: Unukalhai

MONTH WHEN BEST VISIBLE: July – Northern hemisphere

OPHIUCHUS
The Serpent Holder

Rasalhague

M12

M10

Ophiuchus represents the god of medicine. In the sky he is depicted holding a large snake. The constellation contains several globular clusters, the brightest of which are M10 and M12.

DATA FILE

SIZE RANKING: 11th

BRIGHTEST STAR: Rasalhague

MONTH WHEN BEST VISIBLE: July – Southern hemisphere

SCUTUM
The Shield

The northern half of this constellation lies in one of the brightest parts of the Milky Way. It contains the Wild Duck Cluster, or M11, a compact group of 3,000 stars, which is visible through binoculars.

Alpha Scuti

M11

DATA FILE

SIZE RANKING: 84th

BRIGHTEST STAR: Alpha Scuti

MONTH WHEN BEST VISIBLE: August – Southern hemisphere

SAGITTA
The Arrow

This tiny constellation contains only one object of interest: the star cluster M71. The cluster is 12,000 light years away, and an astonishing 10 billion years old.

Gamma Sagittae

M71

DATA FILE

SIZE RANKING: 86th

BRIGHTEST STAR: Gamma Sagittae

MONTH WHEN BEST VISIBLE: September – Northern hemisphere

AQUILA
The Eagle

Aquila represents a flying eagle. Its brightest star, Altair, is a young white star, just 17 light years away. It spins at a speed of approximately 280 km (174 miles) per second.

DATA FILE

SIZE RANKING: 22nd

BRIGHTEST STAR: Altair

MONTH WHEN BEST VISIBLE: August – Northern hemisphere

Altair

VULPECULA
The Fox

This faint constellation is home to PSR B1919+21 – the first pulsar discovered, in 1967. It spins round once every 1.337 seconds, flashing pulses of radiation.

DATA FILE

SIZE RANKING: 55th

BRIGHTEST STAR: Anser

MONTH WHEN BEST VISIBLE:
September – Northern hemisphere

Anser

DELPHINUS
The Dolphin

Delphinus resembles a leaping dolphin. Nicolaus Venator, an Italian astronomer, named its brightest stars after himself (try spelling the names of the stars backwards).

Sualocin

Rotanev

DATA FILE

SIZE RANKING: 69th

BRIGHTEST STARS: Rotanev and Sualocin

MONTH WHEN BEST VISIBLE:
September – Northern hemisphere

EQUULEUS

The Foal

Equuleus is the second smallest constellation in the night sky, larger than only Crux. Named after the brother or son of Pegasus, the winged horse in Greek mythology, it contains little of interest.

DATA FILE

 SIZE RANKING: 87th

BRIGHTEST STAR: Kitalpha

MONTH WHEN BEST VISIBLE:
September – Northern hemisphere

Kitalpha

PEGASUS

The Winged Horse

This constellation rose to fame in 1995 when one of its stars – 51 Pegasi – became the first Sun-like star found beyond the Solar System to have a planet circling it.

DATA FILE

SIZE RANKING: 7th

BRIGHTEST STARS: Enif and Scheat

MONTH WHEN BEST VISIBLE:
October – Northern hemisphere

Scheat

51 Pegasi

Enif

AQUARIUS
The Water Carrier

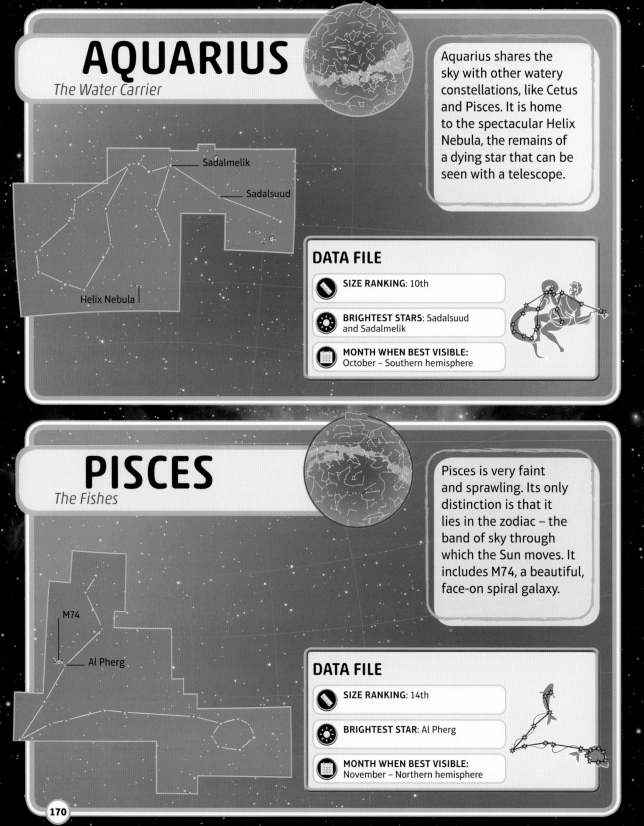

Aquarius shares the sky with other watery constellations, like Cetus and Pisces. It is home to the spectacular Helix Nebula, the remains of a dying star that can be seen with a telescope.

Sadalmelik

Sadalsuud

Helix Nebula

DATA FILE

SIZE RANKING: 10th

BRIGHTEST STARS: Sadalsuud and Sadalmelik

MONTH WHEN BEST VISIBLE: October – Southern hemisphere

PISCES
The Fishes

Pisces is very faint and sprawling. Its only distinction is that it lies in the zodiac – the band of sky through which the Sun moves. It includes M74, a beautiful, face-on spiral galaxy.

M74

Al Pherg

DATA FILE

SIZE RANKING: 14th

BRIGHTEST STAR: Al Pherg

MONTH WHEN BEST VISIBLE: November – Northern hemisphere

CETUS
The Sea Monster

The celestial sea monster Cetus contains a remarkable star, Mira. This bloated red giant swells and shrinks, changing in brightness. At its brightest, it is easily visible to the naked eye.

DATA FILE

- **SIZE RANKING:** 4th
- **BRIGHTEST STAR:** Diphda
- **MONTH WHEN BEST VISIBLE:** December – Southern hemisphere

Mira

Diphda

ORION
The Hunter

Orion is the most spectacular constellation in the sky. The three stars that make up the hunter's belt make it easy to identify.

DATA FILE

- **SIZE RANKING:** 26th
- **BRIGHTEST STARS:** Rigel and Betelgeuse
- **MONTH WHEN BEST VISIBLE:** January – Northern hemisphere

Betelgeuse

Rigel

CANIS MAJOR
The Large Dog

This constellation contains Sirius, or the "Dog Star", the brightest star in the night sky. It is not actually that bright, but only appears so because it lies nearby – only 8.6 light years away.

DATA FILE

⬤ **SIZE RANKING:** 43rd

⬤ **BRIGHTEST STAR:** Sirius

⬤ **MONTH WHEN BEST VISIBLE:**
February – Southern hemisphere

Sirius

CANIS MINOR
The Little Dog

The constellation has one bright star: Procyon. In Greek, its name means "before the dog", because its rising heralded the appearance of Sirius.

DATA FILE

⬤ **SIZE RANKING:** 71st

⬤ **BRIGHTEST STAR:** Procyon

⬤ **MONTH WHEN BEST VISIBLE:**
March – Northern hemisphere

Procyon

MONOCEROS
The Unicorn

Monoceros depicts the one-horned, mythical beast called the Unicorn. The constellation has no bright stars, but it does contain the glorious Rosette Nebula, best seen on long-exposure images.

Rosette Nebula

Alpha Monocerotis

DATA FILE

SIZE RANKING: 35th

BRIGHTEST STAR: Alpha Monocerotis

MONTH WHEN BEST VISIBLE:
February – Southern hemisphere

HYDRA
The Water Snake

Alphard

M83

Hydra is the biggest constellation, stretching more than a quarter of the way around the sky. Of note is M83, a fantastic spiral galaxy that is visible face-on from Earth.

DATA FILE

SIZE RANKING: 1st

BRIGHTEST STAR: Alphard

MONTH WHEN BEST VISIBLE: April – Southern hemisphere

ANTLIA
The Air Pump

Created in the 1750s by the French astronomer Nicolas Louis de Lacaille, this faint constellation contains several double stars, which you can see with binoculars or through a small telescope.

Alpha Antliae

DATA FILE

SIZE RANKING: 62nd

BRIGHTEST STAR: Alpha Antliae

MONTH WHEN BEST VISIBLE: April – Southern hemisphere

SEXTANS
The Sextant

After their observatory burnt down in 1679, Polish astronomers Johannes and Elisabeth Hevelius honoured the instrument for measuring star positions, the sextant, by creating Sextans in 1687.

Alpha Sextantis

DATA FILE

SIZE RANKING: 47th

BRIGHTEST STAR: Alpha Sextantis

MONTH WHEN BEST VISIBLE: April – Southern hemisphere

CRATER
The Cup

Delta Crateris

This constellation represents the cup of water carried to the god Apollo by the crow (Corvus) in Greek mythology. According to the myth, Apollo threw both into the sky.

DATA FILE

- **SIZE RANKING:** 53rd
- **BRIGHTEST STAR:** Delta Crateris
- **MONTH WHEN BEST VISIBLE:** April – Southern hemisphere

CORVUS
The Crow

Though it is small, this star pattern, which represents a celestial bird, has quite a long history – right back to the Babylonians in 1100 BCE, who saw this constellation as a raven.

Gienah

DATA FILE

- **SIZE RANKING:** 70th
- **BRIGHTEST STAR:** Gienah
- **MONTH WHEN BEST VISIBLE:** May – Southern hemisphere

CENTAURUS

The Centaur

This constellation contains Alpha Centauri, the Sun's nearest bright neighbour. Its fainter companion, Proxima Centauri, lies slightly closer to us, at a distance of 4.24 light years.

DATA FILE

- **SIZE RANKING**: 9th

- **BRIGHTEST STARS**: Alpha Centauri and Hadar

- **MONTH WHEN BEST VISIBLE**: May – Southern hemisphere

Alpha Centauri

Hadar

LUPUS

The Wolf

Lupus represents the wolf, although Greek astronomers imagined it as an animal being held up for sacrifice on an altar by Centaurus, the centaur. It is possible to see some lovely double stars with a telescope.

Alpha Lupi

DATA FILE

- **SIZE RANKING**: 46th

- **BRIGHTEST STAR**: Alpha Lupi

- **MONTH WHEN BEST VISIBLE**: June – Southern hemisphere

SAGITTARIUS
The Archer

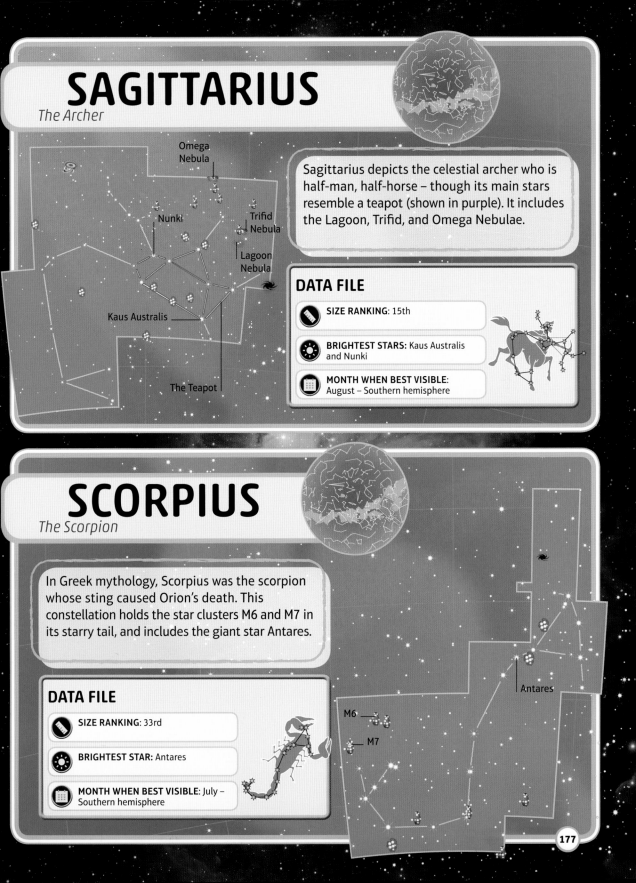

Omega Nebula

Nunki

Trifid Nebula

Lagoon Nebula

Kaus Australis

The Teapot

Sagittarius depicts the celestial archer who is half-man, half-horse – though its main stars resemble a teapot (shown in purple). It includes the Lagoon, Trifid, and Omega Nebulae.

DATA FILE

SIZE RANKING: 15th

BRIGHTEST STARS: Kaus Australis and Nunki

MONTH WHEN BEST VISIBLE: August – Southern hemisphere

SCORPIUS
The Scorpion

In Greek mythology, Scorpius was the scorpion whose sting caused Orion's death. This constellation holds the star clusters M6 and M7 in its starry tail, and includes the giant star Antares.

DATA FILE

SIZE RANKING: 33rd

BRIGHTEST STAR: Antares

MONTH WHEN BEST VISIBLE: July – Southern hemisphere

Antares

M6

M7

CAPRICORNUS
The Sea Goat

This constellation is shown as a goat with a fish tail. A careful look at Algedi reveals it is a pair of stars: each of these is a close double (telescope needed). Binoculars show Dabih is also a double star.

Algedi

Dabih

DATA FILE

SIZE RANKING: 40th

BRIGHTEST STARS: Algedi and Dabih

MONTH WHEN BEST VISIBLE:
September – Southern hemisphere

MICROSCOPIUM
The Microscope

French astronomer Nicolas Louis de Lacaille named Microscopium in the 1750s. He created 14 new constellations that depicted many scientific instruments, here the microscope.

Gamma Microscopii

DATA FILE

SIZE RANKING: 66th

BRIGHTEST STAR: Gamma Microscopii

MONTH WHEN BEST VISIBLE:
September – Southern hemisphere

PISCIS AUSTRINUS
The Southern Fish

This southern constellation contains a bright star called Fomalhaut, which is Arabic for "fish's mouth". This star has a dusty disc, in which the Hubble Space Telescope has found a large planet.

Fomalhaut

DATA FILE

SIZE RANKING: 60th

BRIGHTEST STAR: Fomalhaut

MONTH WHEN BEST VISIBLE:
October – Southern hemisphere

SCULPTOR
The Sculptor

The jewel in this faint constellation is the Silver Coin Galaxy (NGC 253), discovered by astronomer Caroline Herschel in 1783. It is visible, nearly edge-on, through binoculars and small telescopes.

Silver Coin Galaxy

Alpha Sculptoris

DATA FILE

SIZE RANKING: 36th

BRIGHTEST STAR: Alpha Sculptoris

MONTH WHEN BEST VISIBLE:
November – Southern hemisphere

FORNAX
The Furnace

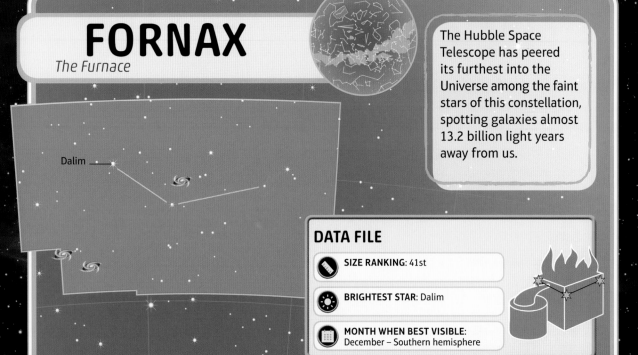

Dalim

The Hubble Space Telescope has peered its furthest into the Universe among the faint stars of this constellation, spotting galaxies almost 13.2 billion light years away from us.

DATA FILE

- **SIZE RANKING:** 41st
- **BRIGHTEST STAR:** Dalim
- **MONTH WHEN BEST VISIBLE:** December – Southern hemisphere

CAELUM
The Chisel

This dim star pattern is one of several 18th-century additions by Nicolas Louis de Lacaille that depict artists' tools. In this instance, it represents a chisel used by engravers.

Alpha Caeli

DATA FILE

- **SIZE RANKING:** 81st
- **BRIGHTEST STAR:** Alpha Caeli
- **MONTH WHEN BEST VISIBLE:** January – Southern hemisphere

ERIDANUS
The River

Eridanus represents a winding starry river, which flows from the bottom of Orion down to the star Achernar. Of note is a star called Epsilon Eridani, which is forming a family of planets.

DATA FILE

SIZE RANKING: 6th

BRIGHTEST STAR: Achernar

MONTH WHEN BEST VISIBLE:
December – Southern hemisphere

Epsilon
Eridani

Achernar

LEPUS
The Hare

This constellation depicts the favourite quarry of the celestial hunter Orion – a hare that crouches at Orion's feet. Lepus contains a star named Arneb, which means "hare" in Arabic.

DATA FILE

SIZE RANKING: 51st

BRIGHTEST STAR: Arneb

MONTH WHEN BEST VISIBLE:
February – Southern hemisphere

Arneb

COLUMBA
The Dove

Dutch map-maker Petrus Plancius created Columba in 1592. It depicts the dove that Noah released from the Ark to find dry land. Its brightest star is Phact, from the Arabic "ring dove".

DATA FILE

SIZE RANKING: 54th

BRIGHTEST STAR: Phact

MONTH WHEN BEST VISIBLE:
February – Southern hemisphere

Phact

PYXIS
The Compass

French astronomer Nicolas Louis de Lacaille devised this faint southern constellation in the 1750s during his survey of the southern sky. Pyxis depicts a magnetic compass, as used on ships.

DATA FILE

SIZE RANKING: 65th

BRIGHTEST STAR: Alpha Pyxidis

MONTH WHEN BEST VISIBLE:
March – Southern hemisphere

Alpha Pyxidis

PUPPIS
The Stern

Puppis contains a major star, Naos, which at 42,000°C (75,630°F) is among the hottest known. This constellation includes several bright star clusters, such as M46 and M47.

DATA FILE

SIZE RANKING: 20th

BRIGHTEST STAR: Naos

MONTH WHEN BEST VISIBLE:
March – Southern hemisphere

M47

M46

Naos

VELA
The Sails

Lying in the Milky Way, Vela contains a treasure trove of stars, star clusters, and nebulae. Binoculars show Suhail to be a lovely double star. The bright star cluster IC 2391 is visible to the naked eye.

Suhail

IC 2391

DATA FILE

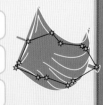

SIZE RANKING: 32nd

BRIGHTEST STAR: Suhail

MONTH WHEN BEST VISIBLE:
March – Southern hemisphere

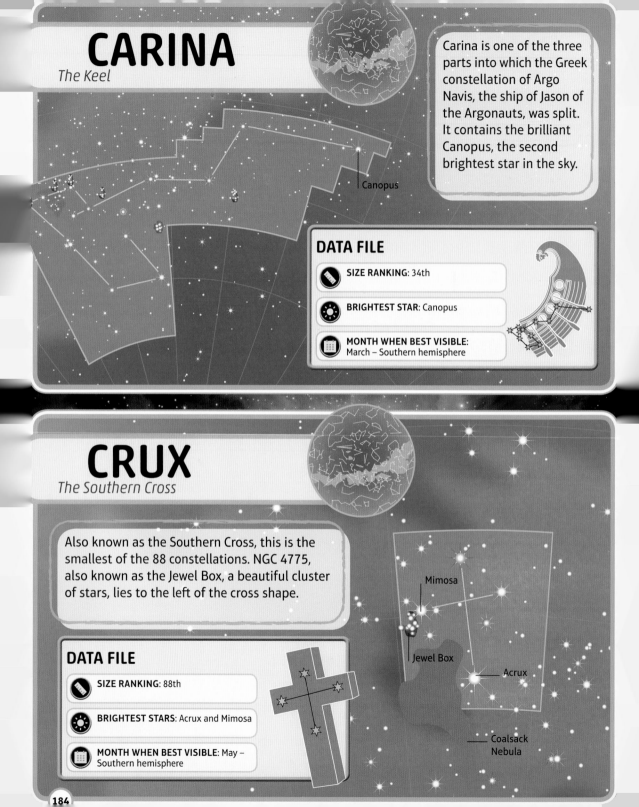

CARINA
The Keel

Canopus

Carina is one of the three parts into which the Greek constellation of Argo Navis, the ship of Jason of the Argonauts, was split. It contains the brilliant Canopus, the second brightest star in the sky.

DATA FILE

SIZE RANKING: 34th

BRIGHTEST STAR: Canopus

MONTH WHEN BEST VISIBLE:
March – Southern hemisphere

CRUX
The Southern Cross

Also known as the Southern Cross, this is the smallest of the 88 constellations. NGC 4775, also known as the Jewel Box, a beautiful cluster of stars, lies to the left of the cross shape.

Mimosa

Jewel Box

Acrux

Coalsack
Nebula

DATA FILE

SIZE RANKING: 88th

BRIGHTEST STARS: Acrux and Mimosa

MONTH WHEN BEST VISIBLE: May –
Southern hemisphere

MUSCA
The Fly

The constellations Musca and Crux share the distinctive Coalsack Nebula – a large, dark cloud in space that is silhouetted against the distant stars of the Milky Way.

Coalsack Nebula

Alpha Muscae

DATA FILE

SIZE RANKING: 77th

BRIGHTEST STAR: Alpha Muscae

MONTH WHEN BEST VISIBLE: May – Southern hemisphere

CIRCINUS
The Compasses

This faint constellation hosted the first known supernova recorded by the ancient Chinese in 185 CE. This supernova, or exploding star, was brighter than Venus in the night sky.

Alpha Circini

DATA FILE

SIZE RANKING: 85th

BRIGHTEST STAR: Alpha Circini

MONTH WHEN BEST VISIBLE: June – Southern hemisphere

NORMA

The Set Square

Lying in the Milky Way, Norma contains several star clusters. With the naked eye you can spot the cluster NGC 6087, centred on a yellow supergiant that pulsates.

DATA FILE

SIZE RANKING: 74th

BRIGHTEST STAR: Gamma Normae

MONTH WHEN BEST VISIBLE: July – Southern hemisphere

Gamma Normae

NGC 6087

TRIANGULUM AUSTRALE

The Southern Triangle

One of the most distinctive constellations near the south pole, the southern triangle has three stars of almost equal brightness marking the corners of an equal-sided triangle.

Atria

DATA FILE

SIZE RANKING: 83rd

BRIGHTEST STAR: Atria

MONTH WHEN BEST VISIBLE: July – Southern hemisphere

ARA
The Altar

Named after the mythological altar on which the gods of Mount Olympus swore an oath of loyalty before fighting the Titans, Ara contains the remarkable NGC 6193, a cluster rich in young stars.

NGC 6193

Beta Arae

DATA FILE

SIZE RANKING: 63rd

BRIGHTEST STAR: Beta Arae

MONTH WHEN BEST VISIBLE: July – Southern hemisphere

CORONA AUSTRALIS
The Southern Crown

A distinct curve of faint stars, Corona Australis represents a wreath of green leaves. The globular cluster NGC 6541 is an interesting object that can be seen with a small telescope.

Alphecca Meridiana

NGC 6541

DATA FILE

SIZE RANKING: 80th

BRIGHTEST STAR: Alphecca Meridiana

MONTH WHEN BEST VISIBLE: August – Southern hemisphere

TELESCOPIUM
The Telescope

French astronomer Nicolas Louis de Lacaille created this faint constellation in the 1750s. The globular cluster NGC 6584 can be seen through a large telescope.

DATA FILE

- **SIZE RANKING:** 57th
- **BRIGHTEST STAR:** Alpha Telescopii
- **MONTH WHEN BEST VISIBLE:** August – Southern hemisphere

Alpha Telescopii

NGC 6584

INDUS
The Indian

This star pattern represents a native hunter brandishing a spear and commemorates the people who lived around the Indian Ocean when it was first explored by Europeans.

DATA FILE

- **SIZE RANKING:** 49th
- **BRIGHTEST STAR:** Alpha Indi
- **MONTH WHEN BEST VISIBLE:** September – Southern hemisphere

Alpha Indi

GRUS
The Crane

The constellation Grus represents a crane, a long-legged wading bird. If you look carefully you can see that several of the stars in this pattern are actually close double stars.

Alnair

DATA FILE

SIZE RANKING: 45th

BRIGHTEST STAR: Alnair

MONTH WHEN BEST VISIBLE:
October – Southern hemisphere

PHOENIX
The Phoenix

Representing a mythical bird that was said to be reborn from its own ashes every 500 years, Phoenix contains Zeta Phoenicis, a star that dims every 1.7 days as a fainter star moves in front of it.

Ankaa

Zeta Phoenicis

DATA FILE

SIZE RANKING: 37th

BRIGHTEST STAR: Ankaa

MONTH WHEN BEST VISIBLE:
November – Southern hemisphere

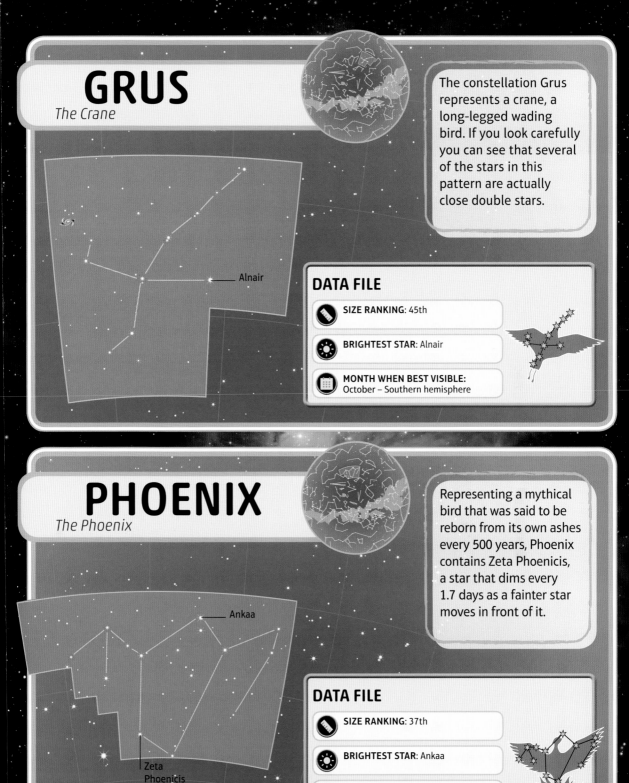

TUCANA
The Toucan

Tucana contains the giant globular star cluster 47 Tucanae and the Small Magellanic Cloud, a mini-galaxy about 200,000 light years away from Earth. Both are visible to the naked eye.

Alpha Tucanae

47 Tucanae

Small Magellanic Cloud

DATA FILE

SIZE RANKING: 48th

BRIGHTEST STAR: Alpha Tucanae

MONTH WHEN BEST VISIBLE:
November – Southern hemisphere

HYDRUS
The Little Water Snake

Created by Dutch explorers in 1597, this constellation is insignificant compared to the other Water Snake, the ancient Hydra. It slithers between the Large and Small Magellanic Clouds.

Small Magellanic Cloud

Large Magellanic Cloud

Beta Hydri

DATA FILE

SIZE RANKING: 61st

BRIGHTEST STAR: Beta Hydri

MONTH WHEN BEST VISIBLE:
December – Southern hemisphere

HOROLOGIUM
The Pendulum Clock

This constellation was named after the clock that French astronomer Nicolas Louis de Lacaille used to keep time in his observatory, when charting the southern stars.

DATA FILE

SIZE RANKING: 58th

BRIGHTEST STAR: Alpha Horologii

MONTH WHEN BEST VISIBLE:
December – Southern hemisphere

Alpha Horologii

RETICULUM
The Net

Originally called the Rhombus for its diamond shape, the Net depicts the cross-hairs in a telescope's eyepiece that lets astronomers pinpoint stars' positions.

DATA FILE

SIZE RANKING: 82nd

BRIGHTEST STAR: Alpha Reticuli

MONTH WHEN BEST VISIBLE:
January – Southern hemisphere

Alpha Reticuli

PICTOR
The Painter's Easel

The second brightest star in Pictor, Beta Pictoris, hit the headlines in 1984: it was the first star discovered to be surrounded by a dusty disc, in which new planets are being born.

DATA FILE

SIZE RANKING: 59th

BRIGHTEST STAR: Alpha Pictoris

MONTH WHEN BEST VISIBLE:
February – Southern hemisphere

Beta Pictoris

Alpha Pictoris

DORADO
The Goldfish

This constellation contains a treasure-trove of nebulae and star clusters within the Large Magellanic Cloud, the nearest major galaxy to the Milky Way.

DATA FILE

SIZE RANKING: 72nd

BRIGHTEST STAR: Alpha Doradus

MONTH WHEN BEST VISIBLE:
January – Southern hemisphere

Alpha Doradus

Large Magellanic Cloud

VOLANS
The Flying Fish

Volans was invented in the 16th century by Dutch explorers and depicts a flying fish – an exotic creature they saw on their voyages. It contains two double stars that can be admired in a telescope.

Beta Volantis

DATA FILE

- **SIZE RANKING:** 76th
- **BRIGHTEST STAR:** Beta Volantis
- **MONTH WHEN BEST VISIBLE:** March – Southern hemisphere

MENSA
The Table Mountain

Astronomer Nicolas Louis de Lacaille named this constellation Mensa, which means "table" in Latin, to honour Table Mountain in South Africa. He studied the stars near the mountain in the 1750s.

DATA FILE

- **SIZE RANKING:** 75th
- **BRIGHTEST STAR:** Alpha Mensae
- **MONTH WHEN BEST VISIBLE:** January – Southern hemisphere

Alpha Mensae

CHAMAELEON
The Chamaeleon

Chamaeleon is a small southern constellation invented in the late 16th century. It is named after the lizard that can change its skin colour to match its surroundings.

Alpha Chamaeleontis

DATA FILE

SIZE RANKING: 79th

BRIGHTEST STAR:
Alpha Chamaeleontis

MONTH WHEN BEST VISIBLE: April – Southern hemisphere

APUS
The Bird of Paradise

This small constellation's name means "foot-less" in Latin. At one point, naturalists in Europe thought that the beautiful birds-of-paradise, which Apus depicts, had no feet.

Alpha Apodis

DATA FILE

SIZE RANKING: 67th

BRIGHTEST STAR: Alpha Apodis

MONTH WHEN BEST VISIBLE: July – Southern hemisphere

PAVO
The Peacock

Peacock

Formed from stars first observed by Dutch sailors in the 16th century, Pavo depicts a peacock, a bird with a fan-like tail. Its brightest star is nicknamed "Peacock".

DATA FILE

SIZE RANKING: 44th

BRIGHTEST STAR: Peacock

MONTH WHEN BEST VISIBLE:
September – Southern hemisphere

OCTANS
The Octant

Introduced in the 18th century, Octans represents a navigation instrument called an octant. The faint southern Pole Star, Sigma Octantis, lies in Octans, but it is barely visible to the naked eye.

Nu Octantis

Sigma Octantis

DATA FILE

SIZE RANKING: 50th

BRIGHTEST STAR: Nu Octantis

MONTH WHEN BEST VISIBLE:
October – Southern hemisphere

REFERENCE

Have you ever wondered where space missions are launched from and how the exploration of space has unfolded? Find the answers here, along with the most amazing facts about extraordinary things in our Universe – from the speed of light to the number of galaxies out there. A glossary explains many of the terms used in the book.

SPACE EXPLORATION
Timeline

The exploration of space started with the launch of the first satellite, Sputnik 1, in 1957. Within 12 years, people had walked on the Moon. By 2016, we had explored the furthest reaches of our Solar System, and the quest for knowledge is set to continue.

Statue of Yuri Gagarin

18 March 1965: First spacewalk (Alexei Leonov).

4 October 1957: First artificial satellite, Sputnik 1, is launched into space.

Sputnik 1

12 April 1961: First human spaceflight (Yuri Gagarin).

| 1950 | 1952 | 1954 | 1956 | 1958 | 1960 | 1962 | 1964 |

16 June 1963: First woman in space (Valentina Tereshkova).

Opportunity

29 October 1991: Galileo probe makes first fly-by of asteroid.

25 January 2004: Rover lands on Mars to study for signs of past water (Opportunity).

Space dog Laika
3 November 1957: First animal in orbit (the dog Laika).

| 1984 | 1986 | 1988 | 1990 | 1992 | 1994 | 1996 | 1998 |

20 February 1986: First consistently inhabited space station, Mir, is sent into orbit.

24 April 1990: Space shuttle launches Hubble Space Telescope.

Hubble Space Telescope

Bruce McCandless III

7 February 1984: First untethered spacewalk (Bruce McCandless III).

Butterfly Nebula as seen by Hubble

4 December 1998: Unity, the second module of the International Space Station, is launched.

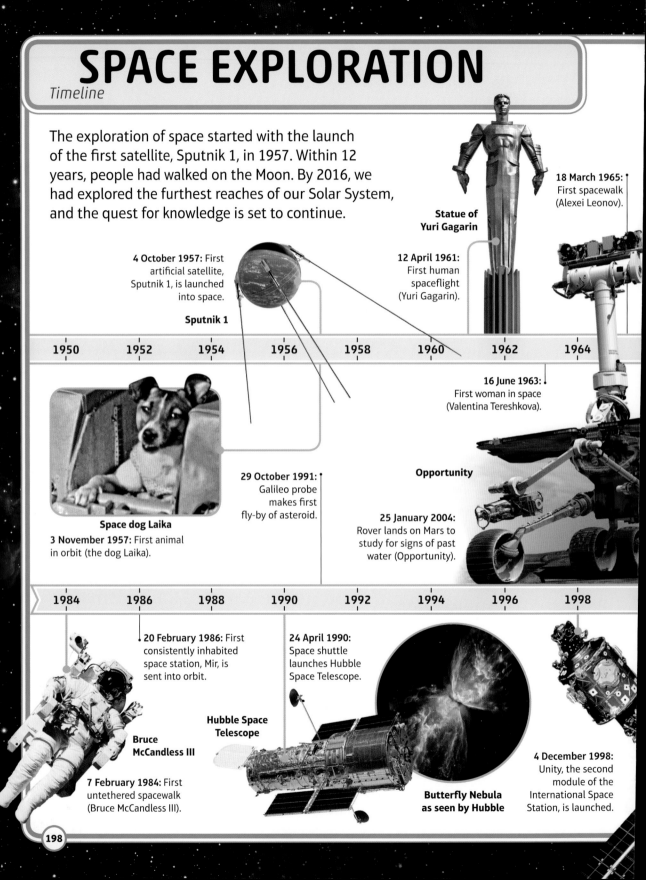

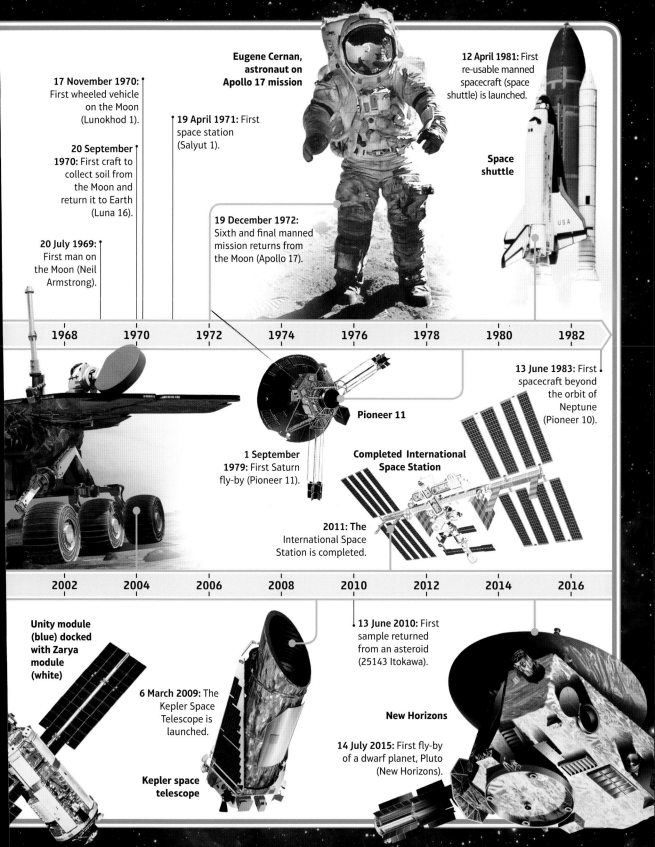

17 November 1970: First wheeled vehicle on the Moon (Lunokhod 1).

20 September 1970: First craft to collect soil from the Moon and return it to Earth (Luna 16).

20 July 1969: First man on the Moon (Neil Armstrong).

Eugene Cernan, astronaut on Apollo 17 mission

19 April 1971: First space station (Salyut 1).

19 December 1972: Sixth and final manned mission returns from the Moon (Apollo 17).

12 April 1981: First re-usable manned spacecraft (space shuttle) is launched.

Space shuttle

1968 1970 1972 1974 1976 1978 1980 1982

13 June 1983: First spacecraft beyond the orbit of Neptune (Pioneer 10).

Pioneer 11

1 September 1979: First Saturn fly-by (Pioneer 11).

Completed International Space Station

2011: The International Space Station is completed.

2002 2004 2006 2008 2010 2012 2014 2016

Unity module (blue) docked with Zarya module (white)

6 March 2009: The Kepler Space Telescope is launched.

13 June 2010: First sample returned from an asteroid (25143 Itokawa).

New Horizons

14 July 2015: First fly-by of a dwarf planet, Pluto (New Horizons).

Kepler space telescope

THE WORLD'S SPACE AGENCIES

Paris

Washington, DC

① ③ ④ ② ⑤ ⑥

NASA

NAME: National Aeronautics and Space Administration

COUNTRY: USA

FOUNDED: 29 July 1958

HEADQUARTERS: Washington, DC

As of 2016, around 70 countries in the world are busy building spacecraft, though only 13 of them have their own rockets to launch spacecraft – including the European Space Agency (ESA), which includes many countries in Europe. Only three of the world's space agencies – NASA (United States), ROSCOSMOS (Russia), and CNSA (China) – have launched humans into space.

ESA

NAME: European Space Agency

COUNTRIES: 22 from Europe, and Canada

FOUNDED: 30 May 1975

HEADQUARTERS: Paris, France

ROSCOSMOS

NAME: Russian Federal Space Agency

COUNTRY: Russia

FOUNDED: 25 February 1992

HEADQUARTERS: Moscow

JAXA

NAME: Japan Aerospace Exploration Agency

COUNTRY: Japan

FOUNDED: 1 October 2003

HEADQUARTERS: Tokyo

CNSA

NAME: China National Space Administration

COUNTRY: China

FOUNDED: 22 April 1993

HEADQUARTERS: Beijing

ISRO

NAME: Indian Space Research Organisation

COUNTRY: India

FOUNDED: 15 August 1969

HEADQUARTERS: Bengaluru

Moscow

Tokyo

Beijing

Bengaluru

KEY LAUNCH SITES

1. Kodiak, USA
2. Wallops, USA
3. Mojave Air and Space Port, USA
4. Cape Canaveral/Kennedy Space Center, USA
5. Kourou, French Guiana
6. Alcantara, Brazil
7. Palmachim, Israel
8. Plesetsk, Russia
9. Kapustin Yar, Russia
10. Baikonur, Kazakhstan
11. Vostochny, Russia
12. Sriharikota, India
13. Jiuquan, Inner Mongolia
14. Xichang, China
15. Taiyuan, China
16. Tanegashima, Japan
17. Uchinoura, Japan

AMAZING SPACE FACTS

Displays in the sky called the Aurora Borealis and Aurora Australis are caused when electrically charged particles from the Sun are trapped by Earth's magnetic poles and collide with gas atoms in the atmosphere over the planet's polar regions. The displays create shifting, coloured light displays in the sky.

We see a total solar eclipse when the Moon precisely covers up the Sun. By an amazing coincidence, the Sun is 400 times bigger than the Moon, and also 400 times further away.

Shooting stars are not stars at all. They are tiny fragments of space dust that burn up in Earth's atmosphere, leaving a trail of light behind them.

If you see a slow-moving dot in the sky, it may be a satellite, lit up by the Sun. You may even spot the International Space Station (ISS).

The furthest object visible to the naked eye is the Triangulum Galaxy, or M33, three million light years away – although you need to be a long way away from lights to be able to see it.

Imagine that you were in a car, travelling at 110 kph (68 mph). To reach the Sun would take 155 years; the nearest star, Proxima Centauri, 2 million years; and the centre of our Milky Way galaxy, 14 billion years.

Space starts only 100 km (62 miles) above our heads – a one-hour drive if your car could go straight upwards.

Pieces of "space junk" – including broken up satellites and tools dropped by astronauts – travel 10 times faster than a bullet, and are very dangerous to spacecraft in space. In 2009, space junk destroyed a communications satellite.

When the Sun dies, five billion years from now, humankind will have to find a new planet circling another star. No problem, there are over 3,000 of them out there.

A British backyard astronomer, Tom Boles, has discovered a record 155 supernovae.

There may be **20 trillion galaxies** in the known Universe.

The world's biggest telescope, the Gran Telescopio Canarias in the Canary Islands, Spain, collects light with a mirror half the size of a tennis court.

Much of the matter in the Universe, including that making up our bodies, is recycled stardust.

Five Solar System planets are easily visible to the naked eye. They are among the brightest objects in the night sky.

Halley's Comet, which visits the Sun every 76 years, was the first comet to be predicted by Newton's law of gravity. Edmond Halley first noticed this, which is why the comet is named after him.

The most brilliant comet ever seen – the Great September Comet of 1882 – was brighter than a full moon.

Space companies are planning to mine asteroids – some contain valuable metals, such as platinum, which could be worth over a trillion dollars.

The biggest crater on Earth, the Vredefort Crater in South Africa, is 300 km (185 miles) across. It was blasted out by an asteroid impact 2 billion years ago.

The difference between a star and a planet when you look into the night sky? Stars are huge nuclear reactors, churning energy into space. Planets are inert, and simply reflect starlight.

Our Milky Way galaxy and the nearby Andromeda Galaxy are approaching each other (unlike most galaxies, which are moving away). In five billion years' time, they will collide, to create a vast galaxy nicknamed Milkomeda.

Life on our planet formed from substances rich in carbon possibly delivered to our planet by comets hitting Earth.

When you see the light from a distant star, it may no longer exist – because its light may have taken thousands of years to reach us.

There are almost a trillion trillion stars in all the galaxies in the Universe. That means there are more stars than there are grains of sand on Earth.

One theory suggests that the Universe may end in a Big Rip: planets will be torn away from stars, then planets themselves ripped apart, and finally all the atoms will explode.

Primitive life, in the form of bugs and bacteria, could exist in many moons in the Solar System. Titan, orbiting Saturn, has seas of ethane and methane: raw materials of life. Jupiter's Europa has an ocean under its icy surface, perhaps teeming with life.

On a clear, moonless night, you can see up to 3,000 stars with the naked eye.

The gravity on the Moon's surface is just 17 per cent that of Earth's. If you jump 1 m (3 ft) high on Earth, you would be able to jump 6 m (19 ft) on the Moon.

Astronauts are up to 5 cm (2 in) taller in space because their spines expand in the weightlessness.

If you counted the stars of the Milky Way at a rate of one a second, it would take you about 5,000 years to count them all.

Yuri Gagarin's mother first found out about his flight into space when news of the mission broke.

Space rockets need to reach a speed of 40,000 kph (25,000 mph) to escape the gravitational pull of Earth. This is called the escape velocity.

Astronauts on the International Space Station have been growing plants in space. The techniques they have developed will be needed on any future mission to Mars, as astronauts will need to grow their own food.

GLOSSARY

Asteroid
A small, irregular Solar System object, made of rock and/or metal, that orbits the Sun.

Asteroid Belt
A doughnut-shaped region of the Solar System, between the orbits of Mars and Jupiter, that contains a large number of orbiting asteroids.

Astronaut
A person who travels to space.

Atmosphere
The layer of gas that surrounds a planet. Also the outermost layer of gas around the Sun or a star.

Atom
The smallest part of a chemical element that can exist on its own.

Big Bang
The explosion that created the Universe billions of years ago; it has been expanding ever since.

Black hole
An object in space with such a strong gravitational pull that nothing, not even light, can escape from it.

Comet
An object made of ice and dust that travels around the Sun in an elliptical orbit. As it gets near the Sun, the ice vaporizes, creating tails of dust and gas.

Constellation
A named area of the sky. The sky is divided into 88 constellations.

Crater
A bowl-shaped depression on the surface of a planet, moon, asteroid, or other body.

Crust
The thin, solid outer layer of a planet or moon.

Dark energy
The energy that scientists believe is responsible for the acceleration in the expansion of the Universe.

Dark matter
Invisible matter that can be detected only by the effect of its gravity.

Dwarf planet
A planet that is big enough to have become spherical but which has not managed to clear all the debris from its orbital path.

Exoplanet
A planet that orbits a star other than the Sun.

Galaxy
A collection of millions or trillions of stars, gas, and dust held together by gravity.

Globular cluster
A ball-shaped cluster of stars that orbits a galaxy.

Gravity
The force that pulls all objects that have mass towards one another. It keeps the planets in orbit around the Sun.

Launch vehicle
A rocket-powered vehicle that is used to send spacecraft or satellites into space.

Light year
The distance travelled by light in one year: 9,460 billion km (5,870 billion miles).

Main sequence
A star that burns hydrogen to helium in its core, a phase that lasts most of its life.

Mantle
A thick layer of hot rock

between the core and crust of a planet or moon.

Matter
Something that exists as a solid, liquid, or gas.

Meteor
A streak of light, also called a shooting star, seen when a chunk of space rock burns up due to friction on entering Earth's atmosphere.

Meteorite
A chunk of space rock that reaches the ground and survives impact.

Milky Way
The barred spiral galaxy that contains the Solar System and is visible to the naked eye as a band of faint light across the night sky.

Module
A portion of a spacecraft.

Nebula
A cloud of gas and dust in space.

Neutron star
A dense, collapsed star that is mainly made of neutrons.

Nuclear fusion
A process in which atomic nuclei join to form a heavier nucleus and release large amounts of energy.

Orbit
The path taken by an object around another when affected by its gravity.

Orbiter
A spacecraft designed to orbit an object but which is not designed to land on it.

Payload
Cargo or equipment, such as a satellite or space telescope, carried into space by a rocket or spacecraft.

Planet
A spherical object that orbits a star and is sufficiently massive to have cleared its orbital path of debris.

Planetary nebula
A glowing cloud of gas ejected by a star at the end of its life.

Probe
An unmanned spacecraft that is designed to explore objects in space and transmit information back to Earth.

Pulsar
A neutron star that sends out beams of radiation as it spins.

Quasar
Short for "quasi-stellar radio source", a quasar is the immensely luminous nucleus of a distant active galaxy with a supermassive black hole at its centre.

Red giant
A large, luminous star with a low surface temperature and a reddish colour, nearing the final stages of its life.

Rover
A vehicle that is driven remotely on the surface of a planet or a moon.

Satellite
An object that orbits a planet.

Soviet Union
A group of countries occupying the northern half of Asia and part of eastern Europe formed after the Russian revolution in 1917. The Soviet Union was dissolved in 1991.

Spacewalk
Activity by an astronaut in space outside a spacecraft, usually to conduct repairs or test equipment.

Star
A huge sphere of glowing gas that generates energy by nuclear fusion in its core.

White dwarf
A small, hot, dense star, the core of a red giant that has shed its atmosphere as a planetary nebula.

INDEX

ACKNOWLEDGMENTS

Dorling Kindersley would like to thank the following people for their assistance with this book:
Agnibesh Das, Antara Moitra, Bharti Bedi, Charvi Arora, Nandini Gupta, Neha Samuel, Rupa Rao, and Tina Jindal for editorial assistance; Namita, Pooja Pipil, Sanjay Chauhan, and Shreya Anand for design assistance; and Helen Peters for the index.

Picture Credits

The publisher would like to thank the following for their kind permission to reproduce their photographs:

(Key: a-above; b-below/bottom; c-centre; f-far; l-left; r-right; t-top)

2 ESA / Hubble: NASA, ESA, and The Hubble Heritage Team STScI / AURA. **4 ESA / Hubble:** ESO and Danny LaCrue (cla). **5 ESO:** A. Duro (cla). **Getty Images:** Bettmann (bc); Imagno (bl). **Science Photo Library:** Emilio Segre Visual Archives / American Institute Of Physics (br). **6 NASA:** (tl). **7 Getty Images:** Elena Pueyo (tl). **8-9 ESA / Hubble:** ESO and Danny LaCrue. **10-201 ESO:** ESO / J. Emerson / VISTA (background). **11 NASA:** WMAP Science Team (tc). **12 Alamy Stock Photo:** Reinhold Wittich / Stocktrek Images (c); Stocktrek Images (cr). **13 ESA / Hubble:** NASA (cb). **Science Photo Library:** NASA / JPL (clb). **14-15 NASA:** X-ray: NASA / CXC / Ecole Polytechnique Federale de Lausanne, Switzerland / D.Harvey & NASA / CXC / Durham Univ / R. Massey; Optical & Lensing Map: NASA, ESA, D. Harvey (Ecole Polytechnique Federale de Lausanne, Switzerland) and R. Massey (Durham University, UK). **16-17 ESA / Hubble:** NASA, ESA, S. Beckwith (STScI), and The Hubble Heritage Team STScI / AURA. **17 ESA / Hubble:** Enrico Luchinat & the ESA / ESO / NASA Photoshop FITS Liberator (cb). **Getty Images:** Robert Gendler / Visuals Unlimited, Inc (crb); Stocktrek Images (bc). **NASA:** NASA / CXC / JPL-Caltech / STScI (br). **18-19 Bill Snyder Astrophotography. 21 ESA / Hubble:** European Space Agency, NASA, G. Piotto (University of Padua, Italy) and A. Sarajedini (University of Florida, USA). (tc). **Science Photo Library:** CELESTIAL IMAGE CO. (tr). **22-23 Science Photo Library:** Mark Garlick. **28-29 Alamy Stock Photo:** Marc Ward / Stocktrek Images. **29 Science Photo Library:** RICHARD BIZLEY (tr). **32-33 Science Photo Library:** DETLEV VAN RAVENSWAAY. **35 BBSO / Big Bear Solar Observatory:** (cra). **36 Science Photo Library:** NASA / Johns Hopkins University Applied Physics Laboratory / Carnegie Institution of Washington (clb). **41 Dreamstime.com:** Deanpictures (cra). **42 NASA:** (bc). **44 Kees Veenenbos** (clb). **46-47 NASA:** NASA / JPL-Caltech. **47 NASA:** NASA / JPL-Caltech / UCLA / MPS / DLR / IDA (tr). **49 NASA:** (tr). **55 Dr Dominic Fortes, UCL:** (tr). **56-57**

Science Photo Library: Lynette Cook. **57 NASA:** (cla); NASA / JHUAPL / SwRI (ca/pluto). **Science Photo Library:** Friedrich Saurer (ca/eris). **58-59 Science Photo Library:** Rev. Ronald Royer. **60 Science Photo Library:** DR SETH SHOSTAK (bc). **62-63 ESO:** A. Duro. **64 Dreamstime.com:** Haris Vythoulkas. **65 Getty Images:** Dan Kitwood (tr); Print Collector. **66 Bridgeman Images:** Iran / Persia: Omar Khayyam as envisaged by an unknown Persian artist, 19th century / Pictures from History. **67 Bridgeman Images:** Germany: A hand-colored plate of the Copernican heliocentric system of the sun and planets in Atlas Coelestis. Johann Doppelmayr, Nuremberg, 1742 / Pictures from History (cl). **Getty Images:** Print Collector. **68 Alamy Stock Photo:** GL Archive; Photo Researchers, Inc (bl). **69 Getty Images:** DEA / G. DAGLI ORTI. **70 Science Photo Library. 71 Dorling Kindersley:** Science Museum, London (br). **Getty Images:** Imagno. **72 Alamy Stock Photo:** North Wind Picture Archives. **Science Photo Library:** Harvard College Observatory (br). **73 Alamy Stock Photo:** Ian Dagnall. **Getty Images:** Florilegius (bl). **74 akg-images:** Universal Images Group / Tass. **75 Alamy Stock Photo:** Granger, NYC.. **76 Alamy Stock Photo:** PF-(bygone1). **77 Getty Images:** Express Newspapers. **78 Alamy Stock Photo:** Sputnik. **79 Getty Images:** Heritage Images. **80 Getty Images:** Bettmann. **81 Alamy Stock Photo:** SPUTNIK. **82 Alamy Stock Photo:** J.R. Bale (cla). **Science Photo Library:** Emilio Segre Visual Archives / American Institute Of Physics. **83 Getty Images:** NASA. **Science Photo Library:** A.sokolov & A.leonov / Asap (br). **84 Science Photo Library:** Emilio Segre Visual Archives / American Institute Of Physics / Science Photo Library. **85 Science Photo Library:** Hencoup Enterprises Ltd; Russell Kightley / Science Photo Library (cra). **86 NASA. 87 NASA. 88 Getty Images:** NASA / Space Frontiers; Stocktrek (bl). **89 Alamy Stock Photo:** epa european pressphoto agency b.v.. **90 Getty Images:** Kimberly White. **91 Getty Images:** Laski Diffusion. **92 Getty Images:** AFP. **Science Photo Library:** Detlev Van Ravenswaay (br). **93 Getty Images:** Saul Loeb. **94-95 NASA. 96 akg-images. Getty Images:** Popperfoto (bl). **97 Science Photo Library:** Detlev Van Ravenswaay. **98 NASA:** (bl). **Science Photo Library:** Detlev Van Ravenswaay. **99 Alamy Stock Photo:** Heritage Image Partnership Ltd (tr). **Science Photo Library:** Detlev Van Ravenswaay. **100 NASA:** JPL (bl); NASA / JPL-Caltech. **101 Alamy Stock Photo:** INTERFOTO / History (bl). **Science Photo Library:** Detlev Van Ravenswaay. **102 NASA. 103 NASA:** NASA Goddard Scientific Visualization Studio / Ernie Wright. **Rex by Shutterstock:** (br). **104-105 NASA. 105 NASA:** (br). **106 NASA:** (bl). **107 Science Photo Library:** Detlev Van Ravenswaay. **108 Getty Images:** SVF2. **109 Science Photo Library:** Sputnik. **110 Getty Images:** Erik Simonsen. **111 Alamy Stock Photo:** NASA / RGB Ventures / SuperStock (br). **NASA. 112 NASA. 113**

Science Photo Library: NASA (cl). **SuperStock:** Science and Society. **114 NASA. 115 Getty Images:** Stocktrek Images. **NASA:** NASA / JPL (br). **116-117 NASA. 118 Science Photo Library:** Mark Williamson. **120 akg-images:** Universal Images Group / Sovfoto. **Science Photo Library:** Sputnik (br). **121 Rex by Shutterstock:** Everett Collection. **122 ESA:** MPAe Lindau (crb). **Science Photo Library:** Detlev Van Ravenswaay. **123 ESA / Hubble:** NASA. **Science Photo Library:** Detlev Van Ravenswaay (crb). **124 Alamy Stock Photo:** Stocktrek Images (crb). **NASA. 125 Getty Images:** VCG. **126 ESA:** ESA–D. Ducros. **127 Science Photo Library:** NASA / JPL; NASA (cla). **129 NASA:** (All Images). **130 NASA:** Chandra X-ray, CXC / NGST (crb); X-ray: NASA / CXC / SAO; Optical: NASA / STScI. **131 NASA:** JPL / JHUAPL (cr); Nssdc / Gsfc (bl). **Science Photo Library:** Johns Hopkins University Applied Physics Laboratory (c). **132 Science Photo Library:** Detlev Van Ravenswaay. **133 NASA:** JPL-Caltech / USGS / Cornell; JPL (br). **134 NASA:** CICLOPS, JPL, ESA, NASA (br). **Science Photo Library:** David Ducros. **135 ESA:** JAXA (cr). **Japan Aerospace Exploration Agency (JAXA):** ISAS / Akihiro Ikeshita. **136 Science Photo Library:** NASA / Ames / JPL-Caltech. **137 NASA:** JPL-Caltech; JPL-Caltech / MSSS (cr). **138 ESA:** ATG medialab; Rosetta / MPS for OSIRIS Team MPS / UPD / LAM / IAA / SSO / INTA / UPM / DASP / IDA (cl). **139 NASA:** NASA / JHUAPL / SwRI. **Science Photo Library:** Johns Hopkins University Applied Physics Laboratory / Southwest Research Institute (br). **140 SpaceX. 141 Alamy Stock Photo:** Konstantin Shaklein. **142-143 Getty Images:** Elena Pueyo. **196-197 NASA. 198 Dreamstime. com:** Nikolai Sorokin (cra). **NASA:** (bl, bc); NASA, ESA, and the Hubble SM4 ERO Team (bc/nebula). **Rex by Shutterstock:** Sovfoto / Universal Images Group (cl). **Science Photo Library:** Detlev Van Ravenswaay (ca). **198-199 NASA:** (b); JPL / Cornell University (c). **199 Getty Images:** Roger Ressmeyer / Corbis / VCG (tr). **NASA:** (tc, br); NASA Ames (c). **Science Photo Library:** NASA / Ames / JPL-Caltech (bc)

Cover images: Front: 123RF.com: Vadim Sadovski cl; **ESO:** cra, VISTA. / Cambridge Astronomical Survey Unit t; **NASA:** clb; **Back: Dreamstime.com:** Eraxion tr; **Fotolia:** Paul Paladin c; **NASA:** crb; Spine: **123RF.com:** Vadim Sadovski t/ (meteor); **ESO:** t, VISTA. / Cambridge Astronomical Survey Unit (Background); **NASA:** t/ (discovery)

All other images © Dorling Kindersley

For further information see:
www.dkimages.com